RENEWABLE ENERGY: RESEARCH, DEVELOPMENT AND POLICIES

FEDERAL GREEN JOBS EFFORTS

RENEWABLE ENERGY ORIGINS AND ASSESSMENT OF OUTCOMES

RENEWABLE ENERGY: RESEARCH, DEVELOPMENT AND POLICIES

Additional books in this series can be found on Nova's website under the Series tab.

Additional E-books in this series can be found on Nova's website under the E-book tab.

RENEWABLE ENERGY: RESEARCH, DEVELOPMENT AND POLICIES

FEDERAL GREEN JOBS EFFORTS

RENEWABLE ENERGY ORIGINS AND ASSESSMENT OF OUTCOMES

ALECIA GRETTER
EDITOR

New York

For permission to use material from this book please contact us:
Telephone 631-231-7269; Fax 631-231-8175
Web Site: http://www.novapublishers.com

Library of Congress Cataloging-in-Publication Data

ISBN: 978-1-63117-269-4

Published by Nova Science Publishers, Inc. † New York

CONTENTS

PREFACE

This book discusses what "green job" are, and provides highlights on challenges of targeted training programs for emerging industries.

Chapter 1 - In the United States, growing awareness of greenhouse gas (GHG) emissions and the possible implications for global climate change have combined with recent high energy prices and economic uncertainty to rekindle interest in renewable energy. Renewable energy technologies generate electricity from resources such as the sun, wind, or biomass, with essentially no net GHG emissions. President Obama has declared a goal for the United States to become the world's leading exporter of renewable energy technologies, setting out policy objectives for the development of related "green jobs".

Green jobs have often been defined to include (at a minimum) jobs that result directly from renewables for generating electricity and powering vehicles as well as jobs that result directly from achieving greater energy efficiency. Studies of green job creation in the renewable energy industry vary greatly as a result of differences in definitions, assumptions, and methodologies, with the resulting analyses producing wide-ranging estimates of the number of green jobs. Complicating the estimation of the number of green jobs is the absence of an authoritative data source. The North American Industry Classification System disaggregates firms into the categories of hydro, fossil fuel, nuclear and "other" sources. Renewables are part of the "other" category. The Bureau of Labor Statistics has requested funding for FY2010 to develop data on the number of green jobs and their characteristics (e.g., wages, training requirements) by industry and occupation.

Most of the future growth in green jobs is generally envisioned as coming from the growth in deployment of renewable energy technologies. Renewable energy deployment programs from state governments have had a great influence on the existing deployment levels of renewable energy technologies and resultant jobs. Historically, the federal investment in renewable energy technologies in the United States has not been about creating jobs, but was focused on developing the technologies to a point where they are ready for commercialization. The timeframe under consideration is thus important in any discussion of the potential for renewable energy technologies to create jobs, for the technologies are at different stages in their development cycles. It is also important to recognize that as a specific renewable energy technology becomes more efficient, the number of jobs per Megawatt of output is likely to decrease.

A key to maximizing green jobs growth in the United States from renewable energy is the domestic design and manufacture of equipment and components. Given the growing international competition for renewable energy markets and green jobs, policy mechanisms and incentives may be necessary to encourage manufacturers to locate production in the United States. Companies must be capable of competing in the domestic market for renewable energy first and foremost, as the potential growth of U.S. renewable markets already has attracted significant international participation. A renewable energy industry capable of serving the export market may create many more jobs than an industry which only serves domestic needs.

Few doubt the potential of renewable energy to help address climate change concerns; the question is whether the desired benefits merit the investment. But developing the next generation of renewable energy technologies and building an internationally competitive industry may require a significant and sustained national investment. Without it, the majority of the solar panels, wind turbines, and components providing the clean energy of tomorrow may continue to be designed and built by workers overseas.

Chapter 2 - Labor received $500 million from the Recovery Act to help create, better understand, and provide training for jobs within the energy efficiency and renewable energy industries, commonly referred to as "green jobs." Since 2009, Labor has also "greened" existing programs and funded additional green jobs training grants and other efforts.

In light of the amount of funding targeted to green programs within Labor, GAO examined: (1) what is known about the objectives and coordination of Labor's green jobs efforts, (2) what type of green jobs training grantees provided and how selected grantees aligned their training to meet employers' green jobs needs, (3) what is known about program outcomes and what challenges, if any, grantees faced in implementing their programs, and (4) what Labor has done to assist and monitor its green jobs grantees. To conduct this work, GAO reviewed relevant federal laws and regulations; surveyed selected offices within Labor using two questionnaires—one for directly-funded green jobs efforts and one for other efforts; interviewed Labor officials and 11 out of 103 green jobs training grantees; and analyzed relevant Labor documents and data.

In: Federal Green Jobs Efforts
Editor: Alecia Gretter
ISBN: 978-1-63117-269-4

Chapter 1

RENEWABLE ENERGY: A PATHWAY TO GREEN JOBS?*

Richard J. Campbell and Linda Levine

SUMMARY

In the United States, growing awareness of greenhouse gas (GHG) emissions and the possible implications for global climate change have combined with recent high energy prices and economic uncertainty to rekindle interest in renewable energy. Renewable energy technologies generate electricity from resources such as the sun, wind, or biomass, with essentially no net GHG emissions. President Obama has declared a goal for the United States to become the world's leading exporter of renewable energy technologies, setting out policy objectives for the development of related "green jobs".

Green jobs have often been defined to include (at a minimum) jobs that result directly from renewables for generating electricity and powering vehicles as well as jobs that result directly from achieving greater energy efficiency. Studies of green job creation in the renewable energy industry vary greatly as a result of differences in definitions, assumptions, and methodologies, with the resulting analyses producing wide-ranging estimates

---

* This is an edited, reformatted and augmented version of a Congressional Research Service publication, R40833, dated September 24, 2009.

of the number of green jobs. Complicating the estimation of the number of green jobs is the absence of an authoritative data source. The North American Industry Classification System disaggregates firms into the categories of hydro, fossil fuel, nuclear and "other" sources. Renewables are part of the "other" category. The Bureau of Labor Statistics has requested funding for FY2010 to develop data on the number of green jobs and their characteristics (e.g., wages, training requirements) by industry and occupation.

Most of the future growth in green jobs is generally envisioned as coming from the growth in deployment of renewable energy technologies. Renewable energy deployment programs from state governments have had a great influence on the existing deployment levels of renewable energy technologies and resultant jobs. Historically, the federal investment in renewable energy technologies in the United States has not been about creating jobs, but was focused on developing the technologies to a point where they are ready for commercialization. The timeframe under consideration is thus important in any discussion of the potential for renewable energy technologies to create jobs, for the technologies are at different stages in their development cycles. It is also important to recognize that as a specific renewable energy technology becomes more efficient, the number of jobs per Megawatt of output is likely to decrease.

A key to maximizing green jobs growth in the United States from renewable energy is the domestic design and manufacture of equipment and components. Given the growing international competition for renewable energy markets and green jobs, policy mechanisms and incentives may be necessary to encourage manufacturers to locate production in the United States. Companies must be capable of competing in the domestic market for renewable energy first and foremost, as the potential growth of U.S. renewable markets already has attracted significant international participation. A renewable energy industry capable of serving the export market may create many more jobs than an industry which only serves domestic needs.

Few doubt the potential of renewable energy to help address climate change concerns; the question is whether the desired benefits merit the investment. But developing the next generation of renewable energy technologies and building an internationally competitive industry may require a significant and sustained national investment. Without it, the majority of the solar panels, wind turbines, and components providing the clean energy of tomorrow may continue to be designed and built by workers overseas.

RENEWABLE ENERGY, ENERGY EFFICIENCY, AND GREEN JOBS

In the United States, growing awareness of greenhouse gas (GHG) emissions and the possible implications for global climate change have combined with recent high energy prices and economic uncertainty to rekindle interest in renewable energy. Fossil fuels have long met the needs of economic growth, which has been a primary driver of energy demand worldwide. But with half of electric power in the United States currently generated from coal, electric power generation is responsible for 40% of domestic carbon dioxide emissions (the primary anthropogenic GHG), and over one-third of all U.S.GHG emissions.[1] The prospect of federal climate change mitigation policies is seen as spurring the growth of renewable energy in the near future. Renewable energy technologies generate electricity from resources such as the sun, wind, or biomass, with essentially no net GHG emissions.

Approximately 567,000 people are employed in the utilities sector in the United States.[2] With renewable energy technologies accounting for about 10 percent of total U.S. energy production,[3] jobs in renewable energy currently represent a small part of this overall workforce. President Obama has declared a goal for the United States to become the world's leading exporter of renewable energy technologies, setting out policy objectives for the development of related "green jobs"[4] and the investment of $150 billion over 10 years in energy research and development (R&D) for the next generation of energy technologies.[5] The American Recovery and Reinvestment Act of 2009 (ARRA, P.L. 111-5) included more than $60 billion for clean energy investments, including $6.3 billion for state and local efforts in renewable energy and energy efficiency, and $500 million for jobs training to help prepare workers for careers in energy efficiency and renewable energy.

Other drivers are also pushing the growth of renewable energy and energy efficiency. Renewable energy technologies can provide clean electricity to help meet regional air quality standards. Goals of energy independence and energy security can also be advanced if trends in fuel efficiency continue and automotive vehicles become significantly powered by electricity from renewable sources. Additionally, as part of corporate social responsibility, many corporations are embracing sustainability[6] as a core principle and consider "green," renewable energy goals as part of the same concept. Industrial processes and supply-chain operations are being examined and redesigned to key on low carbon emission, environmentally-friendly solutions as part of the corporation's perceived overall esponsibility to its customers and the communities impacted by its operations. If such changes are adopted across the economy and incorporated into business norms, green, sustainable practices will then become standard for business operations.

This chapter will examine the current debate on green jobs, looking at the different renewable energy technologies to provide a context for the discussion of the potential for green jobs and issues associated with maximizing growth in the renewable electricity sector.

GREEN JOBS STUDIES

Current policy discussions of shifting the nation toward greater use of renewable sources of energy production almost inevitably have included the subject of green jobs. Although there is no consensus on the term's meaning, it often has been defined to include at a minimum jobs that result directly from increasing reliance on renewables for generating electricity and powering vehicles as well as jobs that result directly from achieving greater energy efficiency. Less often, green jobs have been defined to also include employment that results directly from reducing and mitigating pollution and from conserving natural resources (e.g., water).[7] To this direct employment (i.e., direct jobs), indirect and induced employment are sometimes added as well.[8]

A variety of researchers, usually with the financial support of advocacy groups, have operationalized the diverse definitions of green jobs in order to estimate the number of green jobs at present or at some time in the future. The estimates are based on varying assumptions about, for example, the percentage of U.S. electricity that will be generated by different combinations of renewable resources in the coming decades or the level of public and private sector investments in various renewable energy sources. Less often, the job estimates are based on firm-specific information about the number of persons who are currently employed producing green goods (e.g., solar panels) or providing green services (e.g., energy audits). As a consequence of differences in definitions, assumptions, and methodologies, analysts have produced widely divergent estimates of current and prospective green jobs.

Complicating the estimation of the number of green jobs is the absence of an authoritative data source. Federal statistical agencies, such as the Bureau of Economic Analysis and the Bureau of Labor Statistics (BLS), categorize the data they collect according to the North American Industry Classification System (NAICS). In the case of the utility industry, for example, NAICS disaggregates firms into the categories of hydro, fossil fuel, nuclear and "other" sources. NAICS does not disaggregate the "other" category into those that use renewable sources of electricity production such as wind, solar and biomass. Because renewables are part of the "other" category, no government series provides data on their output and employment. Analysts consequently have articulated. For its part, the government agency responsible for labor force statistics, BLS, has requested funding for FY2010 to develop data on the number of green jobs and their characteristics (e.g., wages, training requirements) by industry and occupation.[9] Industry statistics are expected to be available in FY2011.

In addition, much of the empirical research has estimated only the numbers of green jobs. Estimation or discussion of the net impact on total U.S. employment generally is lacking. Net estimates take into account jobs throughout the economy eliminated or forgone in addition to those created to advance toward a low-carbon environmentally friendly economy. Similarly, the literature does not take into account factors, such as supply constraints (e.g., a tight labor market) and improvements in labor productivity (i.e., output per hour worked), that could affect attainment of estimated levels of green employment.

A review of several analyses of the number of green jobs associated specifically with the use of renewable energy sources is presented below. The distinctions between definitions, assumptions and methodologies of green job studies are addressed when possible.

(1) *Management Information Services, Inc.* (MISI) prepared the report *Defining, Estimating, and Forecasting the Renewable Energy and Energy Efficiency Industries in the U.S. and in Colorado* for the American Solar Energy Society.

The economic research firm developed the following definition of the renewable energy (RE) "industry":

> an employee working in one of the major RE technologies included in this chapter – wind, photovoltaics, solar thermal, hydroelectric power, geothermal, biomass (ethanol, biodiesel, and biomass power), and fuel cells and hydrogen. In addition, in this study, jobs in RE include persons involved in RE activities in the federal, state, and local governments, universities, trade

and professional associations, NGOs, consultants, investment company analysts, etc.[10]

MISI estimated that there were 217,600 direct jobs in the domestic RE industry in 2007. The great majority (154,100 or 70%) were in the biomass segment, chiefly ethanol and biomass power. MISI further estimated the number of indirect and induced jobs supported by the RE industry to be more than twice the number of direct jobs. Although MISI included induced jobs, it acknowledged that some jobs are less green than others (e.g., "ancillary jobs created across the street from a factory producing solar collectors shortly after it opens, such as a doughnut shop, fast food restaurant, dry cleaner, etc. whose customers are primarily the workers at the renewable energy factory").[11] Based on this broad definition, the report states that there were a total of 503,500 (direct, indirect and induced) jobs in the RE industry in 2007.

MISI took the analysis further by specifying, in vague terms, three scenarios for the future use of RE and projecting associated job creation. The base case is a "business as usual" scenario "based loosely on the EIA [Energy Information Administration] reference case."[12] The base case scenario yields a direct, indirect, and induced job total of 1.3 million in 2030. The moderate scenario, which assumes among other things an increase over the base case in government RE initiatives, produces a total of 2.8 million jobs in 2030. The advanced scenario "pushes the envelope" in terms of what might be economically and technologically feasible as a result of greatly heightened government support for RE through 2030.[13] Under the study's most ambitious assumption of RE utilization almost 20 years in the future, 7.3 million direct, indirect and induced jobs might be created.

The analysis does not estimate the impact of labor market conditions on achieving the job projections. If the unemployment rate is low in 2030, there might not be sufficient numbers of jobless workers to fill the projected RE positions. RE employment would either be less than forecast or the jobs would be filled largely by individuals working elsewhere in the economy. In the latter case, job growth in the RE industry would to some degree come at the expense of employment in other industries. The study also does not estimate the number of jobs that might be forgone or lost due to reduced production of energy from fossil fuel sources.

(2) The *Renewable and Appropriate Energy Laboratory* (RAEL) at the University of California- Berkeley prepared *Putting Renewables to Work: How Many Jobs Can the Clean Energy Industry Generate?*

The researchers developed three scenarios which assume that a 20% Renewable Portfolio Standard (RPS) would be attained by 2020, and that electricity demand in 2020 would remain unchanged from 2002 (with gains in efficiency offsetting the usual 2%-3% annual increase in demand). The difference between the three scenarios is the proportions of the RPS composed of biomass (wood and waste) electricity, wind energy and photovoltaic (PV) solar.

Two additional scenarios were developed in which the 20% of electricity produced by renewables in the first three scenarios is instead produced by different mixes of fossil fuels. All five scenarios are shown in Table 1. RAEL concluded that substituting any one of the three renewable scenarios for either of the fossil fuel scenarios would create more direct and indirect jobs. The results also show that increasing use of renewables could result in a reduction of jobs in the fossil fuel industry. In acknowledgement of this point, the authors note that a study conducted by Tellus Institute and MRG Associates for the World Wide Fund for Nature projected a net loss of 23,900 jobs in coal mining and 61,400 jobs in oil and gas extraction by 2020 due to implementation of various clean energy policies.

There are several ways in which the RAEL and MISI studies differ. The primary difference is one of definition, which causes the job estimates of the RAEL analysis to be considerably lower than those of the MISI analysis. First, RAEL does not include induced employment which some do not regard as green jobs. Second, it excludes hydropower. And third, RAEL focuses on electricity generation alone. In addition, the timeframes of the two studies differ (i.e., 2020 RAEL vs. 2030 MISI), and no growth in demand is assumed by RAEL.

(3) *Global Insight* prepared *Current and Potential Green Jobs in the U.S. Economy* for the U.S. Conference of Mayors. The economic research firm estimated that there were 751,051 green (direct and indirect) jobs in 2006, with the largest number related to renewable power generation (127,246). A green job is defined as:

> any activity that generates electricity using renewable [including hydroelectric] or nuclear fuels, agriculture jobs supplying corn or soy for transportation fuel, manufacturing jobs producing goods used in renewable power generation, equipment dealers and wholesalers specializing in renewable energy or energy-efficiency products, construction and installation of energy and pollution management systems, government administration of

environmental programs, and supporting jobs in the engineering, legal, research and consulting fields.[14]

Table 1. Estimated Employment in 2020 of Five Electricity Generation Scenarios

Scenarios	Average Number of Jobs in Construction, Manufacturing and Installation	Average Number of Jobs in Operations and Maintenance and in Fuel Processing	Average Total Employment
20% renewable portfolio (85% biomass, 14% wind, 1% PV)	52,533	111,136	163,669
20% renewable portfolio (60% biomass, 37% wind, 3% PV)	85,008	91,436	176,444
20% renewable portfolio (40% biomass, 55% wind, 5% PV)	111,879	76,139	188,018
Fossil fuels replacing 20% renewables (50% coal, 50% natural gas)	22,711	63,657	86,369
Fossil fuels replacing 20% renewables (100% natural gas)	22,023	61,964	83,987

Source: Daniel M. Kammen, Kamal Kapadia, and Matthias Fripp, *Putting Renewables to Work*, RAEL Report, University of California-Berkeley, 2004.

Notes: Job estimates for PV, wind and biomass are the average of the high and low employment factors from Renewable Energy Policy Project, *The Work that Goes into Renewable Energy*, 2001; Greenpeace, *2 Million Jobs by 2020: Solar Generation*, 2001; and Greenpeace and European Wind Energy Association and Greenpeace, *Wind Force 12: A Blueprint to Achieve 12% of the World Electricity from Wind Power by 2020*, 2003.

Global Insight next forecast the number of green jobs in the categories of renewable electricity generation, residential and commercial retrofitting, and renewable transportation fuels. The second category, retrofitting, is beyond the scope of this chapter and will not be discussed.

In terms of power generation, Global Insight relied on its Energy Group's forecast of a 30% increase in net electricity generation between 2008 and 2038. It assumed that 40% of net electricity generation in the United States during the 30-year period will come from five sources, with wind comprising 30%; solar, 20%; geothermal, 10%; biomass, 30%; and incremental hydropower, 10%.[15] The analysis used several factors to calculate direct green energy jobs in manufacturing and construction and in operations and maintenance. The factors come from research conducted by other parties,[16] which is also the case in the RAEL report. Unlike that study, the Global Insight analysis includes more sources of renewable energy and assumes that energy demand increases over time.

In terms of renewable transportation fuels, Global Insight used a forecast of total fuel production rising to 142,000 million gallons between 2008 and 2038. It assumed that 30% of the total gasoline and diesel consumed by passenger cars and light trucks in 2038 will come from alternative fuels. The analysis used separate factors to calculate direct jobs in manufacturing and construction needed to build additions to the ethanol and biodiesel infrastructure, and in agriculture to grow the feedstock and operate the facilities to turn it into fuel. The factors come from research conducted by the Renewable Fuels Association.[17]

According to this chapter, there might be 407,200 (direct and indirect) jobs in renewable electric power generation by 2018; 802,000 by 2028; and 1,236,800 by 2038.[18] In renewable transportation fuels, there might be 1,205,700 (direct and indirect) jobs by 2018; 1,437,700 by 2028; and 1,492,000 by 2038. Global Insight does not provide estimates of potentially slower job growth elsewhere in the economy due to competition for workers in future years or of jobs forgone due to the assumed percent increases in electricity generation and transportation fuels from renewables. Global Insight does caution that:

> It is important to recognize these forecast results depend heavily on our chosen scenarios. Altering any of the assumptions regarding the share of electricity to be generated from alternative resources, the extent of retrofitting, or the share of transportation fuels from renewable sources would obviously change the results.[19]

(4) In *Study of the Effects on Employment of Public Aid to Renewable Energy Sources*, Gabriel Calzada of the *King Juan Carlos University* analyzed the net jobs impact of Spain's policy initiatives to increase its capacity to generate electricity by wind, mini-hydro, and PV solar power.

The study reportedly was conducted with support from the Institute for Energy Research.[20] It adopted the direct and indirect job creation estimates for Spain's three main renewable sources of electricity that MITRE, a research firm, previously had developed. Specifically, MITRE estimated that Spain's subsidies created 15,000 jobs in wind power, 4,700 jobs in mini-hydro and 14,500 jobs in PV solar between 2000 and 2008, for a total of 50,200 direct and indirect jobs. The researcher then calculated the total public subsidy that created these green jobs to be 28,671 million Euros, or an average government subsidy per worker added of 571,138 Euros.

> In order to know how many net jobs are destroyed by a green job program for each one that it is intended to create, we use two different methods: with the first, we compare the average amount of capital destruction (the subsidized part of the investment) necessary to create a green job against the average amount of capital that a job requires in the private sector; with the second, we compare the average annual productivity that the subsidy to each green job would have contributed to the economy had it not been consumed in such a way, with the average productivity of labor in the private sector that allows workers to remain employed.[21]

Both methods produced the same outcome, namely, an average of 2.2 jobs were not created elsewhere in Spain's economy for each subsidized job created.

This study takes a much different approach than other job creation analyses by estimating the number of jobs had Spain's government not subsidized the three renewable sources of electricity generation. In developing its estimate, the author relies on average measures of capital and productivity across the entire private sector of the economy – but the cost of job creation can differ substantially from economic activity to another. A better comparison might have been with measures of capital and productivity in other electricity-generating industries (e.g., fossil fuel).

(5) Robert Pollin, James Heintz and Heidi Garrett-Peltier of the *Political Economy Research Institute* at the University of Massachusetts prepared *The Economic Benefits of Investing in Clean Energy* with support from the Center for American Progress.

The study, which provides perhaps the clearest explanation of its assumptions and methodology, estimates the number of jobs per year that might be created as a result of spending and other provisions in the American Recovery and Reinvestment Act of 2009 and a version of the American Clean Energy and Security Act of 2009 which is being considered by Congress.

The analysts calculate that the two measures working together might yield $151 billion per year in new "clean-energy investments" made by the government and private firms, and generate a net annual increase of about 1.7 million direct, indirect and induced jobs. This net employment gain is composed of a gross increase economy-wide of approximately 2.5 million jobs and a gross loss of 795,000 jobs if the entire $151 billion annual investment in clean energy came at the expense of the fossil fuel industry. About one-fourth ($41 billion) of the total spending is estimated to occur in the following renewable energy areas: on-grid renewable electricity production and transmission to residential, commercial, and industrial customers ($30 billion), [22] off-grid renewable electricity generated by solar panels on buildings ($3 billion); [23] off-grid nonelectric renewable energy production by geothermal pumps ($3 billion); and alternative motor fuels such as cellulosic biofuels ($5 billion).[24] Thus, the researchers estimate that the great majority of clean-energy investments prompted by the two pieces of legislation involve energy efficiency activities, which they defined as building retrofits, mass transit/freight rail, and smart grid.

The data in Table 2 below underpin the analysts' conclusion that an expenditure on "clean energy sources" creates more jobs than an equivalent expenditure on fossil fuel energy sources. Estimates of jobs per $1 million of spending on renewable resources of energy production are shown in the bottom third of the table, and estimates for comparable spending on fossil fuel sources of energy production are shown in the top third. For example, the wind energy industry (as defined by the analysts) would generate 9.5 direct and indirect jobs per $1 million invested,[25] while $1 million invested in the oil and gas energy industry (as defined by the analysts) would generate 3.7 direct and indirect jobs.[26]

From the perspective of spending that provides the most "bang for the buck" – a major concern when government tries to mitigate the labor market impact of an economy in recession – the results of this analysis suggest that investment in renewable energy sources is superior to fossil fuel sources in terms of job creation. (Both the American Recovery and Reinvestment Act of 2009 and H.R. 2454, the American Clean Energy and Security Act of 2009, have been characterized by some as "jobs bills.") With an unemployment rate of 9.7% in August 2009 and forecasts that the rate will continue to rise through at least year-end and remain elevated for some time to come, the supply of labor is unlikely in the near-term to prevent reaching the analysts' estimate of green job creation. Once the labor market recovers from the recession, however, attainment of the annual job estimate could become increasingly untenable even if $151 billion per year continued to be invested in clean energy (including renewable) sources.

(6) The *Pew Charitable Trusts and Collaborative Economics*, a policy research firm, developed *The Clean Energy Economy: Repowering Jobs, Businesses and Investments Across America.* Unlike the above-described analyses which largely were based on models of the U.S. economy, the authors of this chapter attempted to identify companies engaged in clean energy activities and the number of clean energy jobs at those individual businesses. Put very succinctly, the researchers used multiple sources of information to develop a database of "clean energy economy" firms in selected industry classification codes. Employment statistics for the individual companies in these industry codes came from the National Establishment Time Series (NETS) database, which is based on Dun & Bradstreet information covering the population of business establishments in the United States.[27]

> Our analysis is conservative relative to other studies because we count actual clean energy economy businesses and jobs rather than entire occupations (such as all jobs in mass transit, or all electricians). For example, our report counts the workers who manufacture hybrid cars and buses, technicians who construct wind turbines, electricians who install solar panels on homes and engineers who research fuel cell technology, but it does not include all auto manufacturers, electricians, technicians and engineers. In addition, we focus exclusively on producers and suppliers in the clean energy economy. We do not count jobs that use these products and services—for example, jobs within utilities responsible for purchasing energy monitoring equipment or the mass transit operations that buy hybrid buses—because data limitations prevented the disaggregation of specific jobs within these types of companies.[28]

The researchers define the clean energy economy as (1) clean energy, (2) energy efficiency, (3) environmentally friendly production, (4) conservation and pollution mitigation, and (5) training and support. The five categories are further divided into 16 segments. In the case of clean energy, the segments are energy generation (e.g., wind, solar, geothermal, biomass, hydro, marine and tidal, hydrogen), energy transmission (e.g., power monitoring and metering services, smart grid), and energy storage (e.g., advanced batteries, fuel cells). Nuclear power is excluded "because of significant, ongoing questions about how and where to safely store its waste."[29]

Table 2. Estimated Employment Effect of Investment in Select Energy Sources (number of jobs per $1 million in output)

Energy Source	Direct Jobs[a]	Indirect Jobs[b]	Direct and Indirect Jobs	Direct, Indirect and Induced Jobs[c]
Fossil fuels				
Oil and natural gas	0.8	2.9	3.7	5.2
Coal	1.9	3.0	4.9	6.9
Energy				
efficiency				
Building retrofits	7.0	4.9	11.9	16.7
Mass transit/freight rail	11.0	4.9	15.9	22.3
Smart grid	4.3	4.6	8.9	12.5
Renewables				
Wind	4.6	4.9	9.5	13.3
Solar	5.4	4.4	9.8	13.7
Biomass	7.4	5.0	12.4	17.4

Source: Robert Pollin, James Heintz, and Heidi Garrett-Peltier, *The Economic Benefits of Investing in Clean Energy*, Political Economy Research Institute, June 2009.

Notes: The definitions of direct, indirect, and induced effects appear on p. 27 of the above-captioned report.

a. "Direct effects. The jobs created by retrofitting homes to make them more energy efficient, or building wind turbines."

b. "Indirect effects. The jobs associated with industries that supply intermediate goods for the building retrofits or wind turbines, such as lumber, steel, and transportation."

c. "Induced effects. The expansion of employment that results when people who are paid in the construction or steel industries spend the money they have earned from produce these immediate and intermediate goods for clean energy industries on other products in the economy."

In 2007, according to the Pew analysis, there were 770,385 jobs in the clean energy economy. The clean energy category of renewable sources of energy generation, transmission and storage accounted for 89,000 jobs or 11.6% of the total. About two-thirds of the jobs (501,551) were in the conservation and pollution mitigation category. Each of the remaining categories accounted for fewer jobs than clean energy: 73,000, energy efficiency (e.g., geothermal heating and cooling, smart lighting); 53,700, environmentally friendly production (e.g., biofuel, coal gasification, ethanol, hybrid vehicles, biodegradable products; and aquaculture); and 50,000, training and support (e.g., research in renewable sources of energy generation and in alternative fuels, emissions trading and offsets, and project financing). Some of the jobs in these three categories likely were included by other analysts in their renewable energy estimates. For example, the MISI analysis appears to define the RE industry to encompass related research and investment activities. The Pew report, in contrast, classifies such activities in its "training and support" rather than "clean energy" category.

In summary, the preceding studies of green job creation in the renewable energy industry vary greatly in a few key respects. As a result of differences in definitions, assumptions, and methodologies, the analyses produce wide-ranging estimates of the number of green jobs.

Renewable Energy Technologies and Green Jobs Growth

As previously noted, there is no universally-accepted definition of green jobs. The preceding studies of green jobs generally consider these to be jobs in energy efficiency activities and/or renewable energy. Improving the energy efficiency of current buildings and residences is seen as the easiest way to reduce GHG emissions, and is an area in which many workers could be employed for years to come in energy audit and retrofit activities.[30] But energy efficiency may see as much or more achieved from new construction standards than may be achieved from retrofits to today's existing buildings and housing.

Most of the future growth in green jobs is generally envisioned as coming from the growth in deployment of renewable energy technologies. A common misperception in the studies of green jobs seems to be that all renewable energy technologies have generally the same qualities, with regard to job creation potential. Little effort is made to differentiate between them.

However, renewable energy technologies are designed to harness renewable energy resources with very different physical characteristics, such as the wind and the sun. Different technologies have seen different levels of investment over the years based upon various evaluations of potential and economic readiness to serve current markets or applications. The timeframe under consideration is important in any discussion of the potential for renewable energy technologies to create jobs, for the technologies are at different stages in their development cycles and have attributes suited to different applications. Consequently, the costs of generating electricity varies with each renewable energy technology, and the potential for deployment often depends upon local incentives and the quality of the renewable resource. As such, renewable energy deployment programs from state governments have had a great influence on the current deployment levels of renewable energy technologies and resultant jobs.[31]

Historically, the federal investment in renewable energy technologies in the United States has not been about creating jobs, but instead simply focused on developing the technologies to a point where they are considered ready for commercialization. Even though renewable energy projects are currently being deployed, further development of the technologies is needed if these technologies are to meet the perceived future needs of the marketplace. Some of today's renewable energy technologies may in fact turn out to be "transitional" as advances in technology have and could further lead to new research directions. Innovation could be stimulated from within the industry or induced by developments ancillary to the industry, such as the development of the Smart Grid[32] or schemes to encourage Demand Response.[33] Further research, development, and deployment is likely necessary to develop a more robust renewable energy industry and potential spillover benefits[34] to the economy at large from the R&D spending.

Enabling mechanisms for renewable energy technologies, such as those discussed to help with deployment (i.e., the feed-in tariff in Europe), were designed not so much to create jobs as they were to encourage the use of the new technologies. The U.S. government is now discussing how to invest in renewable energy technologies as a growth engine for future jobs. The investment may not result in large direct jobs creation in the short-term, but prepares for potential growth in the near future especially as climate change considerations may shift paradigms regarding electricity generation choices. This transition would likely require further domestic policies to develop the supporting infrastructure and supply chains to enable the success of new clean energy industries. Additionally, a transition to low carbon energy technologies may result in jobs being lost in other parts of the economy, but this may be considered part of the opportunity cost of making the decision.

Many of the studies concerning green jobs and renewable energy technologies provide estimates of employment reflecting the state of technology at a specific point in time. It is important to recognize that as a specific renewable energy technology becomes more efficient, the number of jobs per Megawatt (MW) of output is likely to decrease. The increase in efficiency may also be reflective of an increase in productivity of manufacturing processes. This has particular relevance if most jobs in renewable energy are likely to be in manufacturing and construction, as opposed to fossil energy in which most jobs relate to fuel processing and operations and maintenance activities.[35]

Outlook for Renewable Energy in the United States

Renewable energy resources may be defined as naturally replenishing (in that they are virtually inexhaustible) resources limited in the amount of energy that is available in a particular period of time.[36] This definition illustrates both the promise and the problems associated with renewable energy technologies which seek to produce electrical energy from renewable resources.[37] Many of today's renewable energy technologies were largely developed in the United States as the federal government explored ways to reduce our dependence on imported oil. Renewable electricity technologies are designed to harness very different renewable energy resources. Some of these renewable resources are fairly constant and predictable, while others can vary significantly according to location, and even time of the day or year. Estimates of market growth are provided to illustrate the potential size of future markets for a particular renewable energy technology.

The U.S. Department of Energy's (DOE's) Energy Information Administration (EIA) annually performs an analysis of trends to arrive at long-term projections of energy supply, demand, and the effect on energy prices. For 2009, the *Annual Energy Outlook* (AEO 2009) has a Reference Case which assumes that existing laws and regulations as of 2007 will be maintained through to 2030. This scenario expects a global economic recovery to begin in 2010, spurring demand for oil and corresponding increases in oil prices to $110 per barrel (2007 dollars) by 2015, increasing to $130 per barrel in 2030. The AEO 2009 Reference Case therefore does not assume climate change legislation will be in effect for the period.[38]

This can be compared to EIA's case for a future in which GHG mitigation is required by legislation and calls this scenario the *LW110* case, modeling such a provision on S. 2191, the Lieberman-Warner Climate Security Act of 2007 introduced in the 110th Congress. EIA also has a scenario called *No GHG Concern* case which removes cost-of-capital adjustments for GHGintensive technologies from the Reference Case. The AEO 2009 Reference Case shows all renewable energy sources accounting for approximately 8.5% of electricity generation in the base year of 2007.[39] Projecting forward to 2030, the Reference Case sees renewable electricity growing to 14% of total generation. Using the LW1 10 case, the analysis projects renewable energy sources accounting for 22% of all electricity generated.

EIA also has analyzed two additional scenarios specifically for changes in the cost of renewable energy technologies. In the case of High Renewable Energy Cost, capital costs, operating and maintenance costs, and performance levels for wind, solar, biomass, and geothermal resources are assumed to remain at 2009 levels through to 2030. In this case, it is assumed that dedicated energy crops do not become available. In its *Low Renewable Energy Cost* scenario, EIA assumes that levelized[40] costs of generating renewable energy will decline 25% below AEO 2009 Reference Case costs by 2030. Generally, this is realized through lower capital costs to construct the plants, and in the case of biomass, fuel supplies are assumed to be 25% less expensive compared to the Reference Case costs. Under the Low Renewable Energy Cost scenario analysis of AEO 2009, renewable electricity sources could account for as much as 22% of power generation for the (non-nuclear) electric power and end-use sectors by 2030.[41]

What these scenarios illustrate is that domestically, a relatively high rate of growth may be expected for renewable electricity overall even if no federal GHG mitigating legislation is enacted during the period to 2030, based on the conditions described.

Outlook for Renewable Energy Markets Overseas

According to EIA projections in its International Energy Outlook for 2009 (IEO2009), global renewable electricity generation could rise by an average 2.9% per year (from 2006 to 2030). If realized, that would mean that renewable electricity would make up 21% of global supply in 2030 as compared to 19% in 2006.[42] With oil prices expected to increase with an expected global economic recovery, renewables are projected to be fastest-growing source of electricity internationally. Wind and hydroelectric power are expected to represent much of the increase. Solar power can be cost-competitive in areas with especially high electricity prices or where government incentives are available. Government subsidies often provide the necessary support for building renewable generation facilities. Renewable sources also offer an opportunity for electrification of regions with unreliable or no centralized electricity services.

RENEWABLE ENERGY TECHNOLOGY STATUS AND DEVELOPMENT NEEDS

A summary of the current status of current major renewable energy technologies follows, using figures for domestic estimated growth based on AEO 2009 projections. Some of the major perceived barriers the technologies must overcome in order achieve a greater share of the electricity generation market are presented. Capacity estimates are for the electric power sector, whose primary business is to sell electricity to the public. Estimates of net summer capacity in 2030 are from AEO 2009's *Low Price* scenario (which assumes technology improvements and other factors will continue a trend of declining prices for electricity from renewable energy technologies) unless otherwise stated.

Biomass

Biomass for electric power or biopower has a large potential and is arguably the most conventional of all renewable electricity technologies. Biomass is the largest source of renewable energy in the United States. Approximately 53% of all renewable energy comes from biomass, represented by biofuels, landfill gas, biogenic municipal solid waste, wood, wood-derived fuels and other biomass such as switchgrass and poplars. Agricultural wastes (such as corn stover) are another potential feedstock. With wood and biomass net summer capacity reported at 7 Gigawatts (GW) for 2007, DOE estimates that 41 GW of domestic biomass generation could be available by 2030.[43] Sustainable management of biomass resources, especially forests, will be critical to this future.

Biomass combustion is a relatively mature technology but it is not widely used and is generally not very efficient unless it is used in a combined heat and power application. As of 2003, there were approximately 66,000 direct jobs in the biomass power segment.[44] Large scale co-firing of biomass with coal is a higher efficiency, lower per unit cost application. Technologies for biomass gasification could result in higher efficiencies when used to produce synthesis gas or hydrogen for heat and/or power production. Demonstration and deployment of newer industrial gasification technologies is needed to scale-up plants and provide economical designs with high degrees of availability. Huge potential exists for the biomass category, as it is generally regarded as a carbon-neutral source of energy. Wood-burning stoves and solar water heaters are the most common applications of residential renewable energy. But improvements are needed to increase fuel efficiency and lower toxic air pollutants for wood stoves.

Liquid biofuels can readily be made from biomass using fermentation processes. Ethanol, the most widely produced biofuel today, is commonly blended with gasoline as a transportation fuel thus reducing GHG and particulate emissions. Other biofuels with transportation potential are methanol and butanol, both of which can be made from renewable sources such as wood wastes, agricultural or municipal solid wastes. Methanol also has electric power production potential, and can be used to fuel combustion turbines to produce electricity, either directly or in its dehydrated form of DiMethyl Ether, a diesel substitute.

Wind

Electricity produced from wind power is growing at a faster rate than natural gas, with a net summer capacity in 2007 of 16.23 GW.[45] This growth has caused a surge in employment with over 85,000 jobs reported by the American Wind Energy Association, of which about 8,000 jobs are construction-related.[46]

Most wind power projects today are onshore, but several coastal regions around the United States also have good quality offshore wind resources that may see development in the near future. Wind turbine sizes are increasing in generating capacity, with domestic turbines in 2007 averaging 1.6 MW in size.[47] When the wind is blowing at speeds which can be harnessed, electricity can be generated at prices nearly competitive with conventional fossil energy based generation. Most of today's wind turbines use a three-bladed windmill rotor design to generate electricity from the wind, but new turbine designs are being developed to generate power at lower wind speeds. Since the wind doesn't blow all the time and varies in strength, integration of large amounts of wind into an electricity grid has often been raised as an issue.[48] Backup generation in the form of natural gas combustion turbines has been the standby choice in some instances. Energy storage (using batteries or other means) is often suggested as a potential answer to deal with intermittency and load variability concerns. Since many of the best wind resource areas in the United States are far from population centers where the electricity generated will be used, the development of transmission facilities to carry power to population centers will likely be needed for wider development. Given these factors, DOE estimates domestic capacity from wind could reach 61 GW by 2030. Additionally, DOE projects as much as 20% of the nation's electrical supply could be provided by wind energy by 2030. This would require wind power capacity to reach 300 GW, or a growth of over 280 GW over the next 21 years. Achieving such a prodigious goal would mean addressing significant challenges in technology, manufacturing, employment, transmission and grid integration, markets, and siting strategies.[49]

Solar PV

Another renewable resource with enormous potential is sunlight. Solar photovoltaic power converts sunlight directly into electricity using photovoltaic (PV) cells which today are largely made from crystalline silicon. Research is underway to reduce the cost of PV cells using base materials other than silicon (such as cadmium telluride) and improved manufacturing techniques which may increase the efficiency of solar cells.[50] Recent advances in "thin-film" technology using cadmium telluride have resulted in module production costs dropping from $2,940 per kilowatt (kW) of capacity in 2004 to $1,120 per kW in 2007.[51] DOE estimates that there were 46 companies manufacturing PV modules and cells in the United States in 2007, with 6,170 person- years of direct employment (representing the equivalent of 6,170 full-time jobs).[52]

Solar PV is currently used in a number of off-grid applications where distributed energy resources are useful, and in peaking power applications to reduce power usage from the electric utility grid. Battery storage is important to off-grid usage to extend hours of usage past peak daylight. While Solar PV installations only represented 0.47 GW of capacity in 2007, DOE estimates generating capacity from solar PV could reach almost 18 GW by 2030 in the United States.[53]

Since the amount of electricity that can be produced from solar PV depends on the intensity of sunshine (and the angle at which PV panels face the sun), significant potential exists for applications in sunny regions. But the relative success of solar PV installations in Germany (with a capacity of 3,000 MW in 2007) prove the wider applicability of the technology in less than optimal climes.[54] Integration of PV cells and materials into building structures and designs for could be a major step for the technology.

The cost of electricity in the United States today from solar PV is relatively high at approximately 18 to 23 cents per kilowatt-hour (kWh) when compared to electricity from fossil energy sources. However, the costs of solar PV capacity have been decreasing. Increases in the efficiency of solar PV cells, combined with other technological advances, could result in solar PV prices of 5 to 10 cents per kWh by 2015[55] which is close to parity with fossil-fueled electricity.

Concentrating Solar Thermal

Concentrating solar thermal technologies use mirrors to concentrate sunlight and generate heat usually for steam production. This steam is then used to generate electricity or provide high- temperature hot water for industrial or other process uses such as heating and cooling. Fairly large areas of land are needed, plus access to water since it is used for steam generation and condenser cooling purposes. Some novel applications heat air directly to generate thermal gradients which are harnessed to produce electricity. Solar thermal technologies are currently used for utility-scale power generation, but costs of 12 cents per kWh in the United States are higher than for fossil- fuel power generation. Advances in the designs and materials of absorbers, reflectors and heat transfer fluids in next-generation solar thermal systems could potentially reduce costs to 6 cents per kWh by 2015.[56] As of 2007, DOE estimates that employment in the U.S. solar thermal industry represented the equivalent of 686 manufacturing jobs.[57]

Improved energy storage schemes would benefit both conventional power generation and off-grid applications in particular. New utility-scale solar thermal facilities will likely be located mostly in the southwestern region of the United States, an area where water resources may be under stress.[58] As such, EIA projects slow growth in domestic power generation from solar thermal technologies from 0.53 GW capacity in 2007 to 0.86 GW by 2030.

Applications in residential and smaller industrial/commercial facilities are another area of potential solar thermal use. Solar hot water heaters are growing in use in the United States, with wide applications in parts of Europe and Asia. Additional R&D investment may be needed to increase energy conversion efficiencies and bring down energy costs if smaller solar thermal systems are to become mainstream choices domestically.

Geothermal

Steam or hot water extracted from geothermal reservoirs in the Earth's crust can be used to generate electricity, or to provide thermal energy for heating or thermal processes. EIA projects this hydrothermal capacity could reach 3 GW by 2030, up from 2.36 GW in 2007.[59] But geothermal energy may no longer depend upon the availability of suitable natural geothermal resources. Enhanced Geothermal Systems (EGS) are man-made geothermal reservoirs. By drilling into the Earth's crust and injecting water to create steam, EGS offers the potential to provide geothermal energy almost anywhere, not just in areas where steam or hot water occur naturally. EGS offers the possibility for large scale generation of clean energy. Improvements in drilling technology can lower costs, and better fluid flow can increase the amounts of power generated.

Ground Source Heat Pumps (GSHP) are used mostly in residential applications and take advantage of the differential of the ground compared to surface air temperatures. GSHP require a piping network to be buried underground to serve the customer's heating and cooling needs. Energy efficiency standards focused on home heating and cooling could lead to improved technologies and wider deployment in new construction. Direct employment in the geothermal heat pump sector as of 2007 represented the equivalent of 1,219 jobs.[60]

Hydroelectric Power

Only 2,400 of the 80,000 dams in the United States produce electricity.[61] Building a new hydroelectric power plant is expensive, and construction uses much water and land. As such, with most of the better sites already developed, DOE does not expect much growth in large conventional hydroelectric capacity. But DOE has identified approximately another 5,677 sites with the potential to generate about 30 GW of power using small-scale hydroelectric technologies. Opportunities with small head and low flow applications may need further R&D to optimize the power generation potential. Most of the future employment opportunities in hydroelectric power are likely to come from small scale projects.

Hydrokinetic energy technologies generate power from the movement of water. Electricity can be generated from the flow of water in rivers, or additionally from the flow of released water at existing dams. Wave energy and tidal applications are just beginning demonstrations of the various technologies to tap the potential.

Micro- and pico-hydropower are options for power in "village scale" or distributed settings. Transmission lines from turbine to service application may be required. Battery storage may not be necessary, depending on the application. Micro-hydropower systems are less than 100 kW. Pico-hydropower systems are less than 5 kW.

FOCUSING ON POTENTIAL MARKETS FOR JOBS GROWTH

If green jobs from renewable energy technologies is a goal, then understanding the future needs and structures of markets for renewable electricity could help the development of a market focus, and provide a strategy for U.S companies to become a global provider of equipment for the next generation of renewable energy technologies. Serving these demands can potentially mean growth in domestic green jobs. While most studies of the potential for green jobs from renewable energy focus on growth in U.S. electricity markets, the developing world may present opportunities for the future. Evolving domestic energy policy concepts (such as the development of a Smart Grid or Demand Response programs) may have an impact on the physical infrastructure and development of services in energy markets internationally. The sections that follow describe some of the major trends which are likely to influence future renewable electricity markets.

Centralized vs. Distributed Electricity Generation

Most communities in the United States are served by a local electric utility which brings power to businesses and residences via electric transmission lines from steam-electric power generating facilities located miles away from large population centers. In fact, a network of power plants generally connects to a grid of transmission lines which then distribute electric power over smaller lines for customer use. These plants commonly burn coal or natural gas (i.e., fossil fuels) to generate electricity, and are usually operating twenty-four hours a day. Efficiency of operation (i.e., as ratio of fuel input to energy produced) has been achieved from large economies of scale, resulting in electrical power being generated fairly cheaply, but with a consequence of commensurately large GHG emissions. This central station concept allows for a number of power plants to serve multiple communities, providing reliable power, and allowing different generating options to be incorporated into networks including renewable electricity.

Some renewable electricity technologies are also capable of serving as base load capacity, notably geothermal, biomass, and hydroelectric facilities with impoundments. Wind power is considered intermittent because the wind doesn't blow all the time, and may not be predictable as a resource. Similarly, solar facilities are dependent on sunlight. Storage schemes may provide an answer for variability of power generation issues thus improving dispatchability, i.e., the capability to generate power to meet system loads. This could allow for more widespread use of renewable energy technologies for traditional electric utility-type operations. Energy storage capabilities could also be key to increasing distributed generation applications of renewable electricity technologies.

Distributed generation does not rely upon a grid or centralized power station for electricity, instead allowing electric power to be generated at or near the point of consumption (i.e., the customer or load). Wind and Solar PV are well-suited to distributed generation, serving as backup or main power supplies for single customers or communities. Similarly, on-site generation is used by industrial facilities such as pulp and paper mills with biomass generation. Communities or consumers seeking back-up or independent power supplies using stand-alone generators or renewable electricity technologies may seek to take advantage of these opportunities.

Some renewable electricity technologies lend themselves to modular, scalable *energy* solutions which may be more favorable to distributed energy solutions. While traditional lead-acid batteries are usually thought of for today's electricity bulk storage systems, new energy storage technologies are on the horizon. For example, advanced battery and fuel cell technologies may be able to efficiently use hydrogen from dissociated water, thus employing distributed generation technologies like solar PV or wind power to generate hydrogen as well as power during peak hours of operation, and provide power for night-time use. Charging stations for electric vehicles may also be ideal applications for renewable electricity solutions.

Developed vs. Developing Economies: A Two-Tiered Energy Future?

In 2009, developing countries surpassed developed countries in total energy use for the first time, reaching a high of 51.2%.[62] EIA has estimated a growth rate for renewable electricity technologies in non-OECD[63] countries at 3.2% annually for the period 2006 to 2030, compared to 2.5% for OECD countries in the same period. Countries in the developed world may be moving towards building economies based on a low-carbon, technology-based energy future following legislative proposals for climate change mitigation. More robust versions of today's renewable electricity technologies could be the future, hybridized with other renewable technologies or possibly natural gas-powered fuel cells or generation. The evolution of a Smart Grid could enable wider deployment of renewable energy technologies as part of a larger centralized network, or alternatively, it could provide for more distributed customer based applications interfaced with electric utility controllers allowing for Demand-Side Management options.[64]

In contrast, the cost and refinements of a smart grid future may not be necessary at this time for markets in the developing world. But that does not mean that these markets do not provide opportunities for U.S. firms. Many of today's existing renewable energy technologies may be adequate for many "village" or distributed applications. Grid-connected electricity systems in many developing countries may not be as reliable or robust as in the United States, providing an opportunity for U.S. manufacturers to provide modular renewable systems as "back-up" or alternative power systems in areas of high-electricity cost. Energy-efficient appliances specifically designed for renewable-powered village applications in developing countries is another possible area for U.S. manufacturers to focus on product development.

Renewable Energy and Climate Change Assistance

Developing countries are less likely to have the resources to afford technologies to help mitigate the potentially detrimental effects of climate change, therefore providing assistance for developing countries to deploy clean energy technologies has been discussed at international forums.[65] By understanding the needs and operational parameters of segments of these markets, or designing products to serve the specific needs of communities in developing countries, U.S. manufacturers could build products for markets in almost any country. Distributed generation schemes can bring renewable sources of electricity to off-grid communities (or communities underserved by a central grid) in developing countries. Such "village applications" could serve GHG reduction goals, enable access to equipment and services which could help improve the health of local populations, and further goals of good will. Electrification is often the first step to the development of modern services for many regions and an educational system providing access to centralized teaching and tools.

By providing options or simply giving assistance in the form of renewable electricity technologies, developed economies can help developing countries avoid GHG emissions from coal or other fossil fuel power generation technologies which may be easier to acquire. U.S. legislative proposals for climate change mitigation include provisions for use of *emissions allowances* to provide exactly such assistance.[66] In the short-term, U.S.-made products could be provided to communities in developing countries. A possible positive benefit of such assistance could be the development of "brand awareness," which may help to develop long-term customers interested in purchasing U.S. manufactured products.

DEVELOPING AN INDUSTRY, NOT JUST JOBS

Investment in renewable energy technologies in the United States until recently was not about creating jobs. Any employment in areas related to or growing out of renewable energy has often been considered as a spillover economic benefit of the original R&D investment rather than a specific focus. Helping technology ventures to become viable manufacturing companies capable of competing for the domestic clean energy market is another step. Supporting structures involving academic institutions and manufacturing supply chains may have to be developed to enable a larger, competitive, domestic renewable energy manufacturing capability.

Transitioning renewable energy companies into an industry capable of competing for global clean energy markets to build related green jobs in the United States over the longer-term may require coordinated policies and further government involvement. Companies must be capable of competing in the domestic market for renewable energy first and foremost, as the potential growth of U.S. renewable markets already has significant international participation. Success in building new manufacturing industries can often be traced to planned, cooperative approaches working with government entities. States and local communities have embraced these ideas for years as part of public-private partnerships in assisting technology to manufacturing ventures.

Globally Competitive Manufacturing Clusters

Many economists believe that complex industrial ecosystems, or *Clusters*, are the drivers of a growing economy. [67] Clusters can be defined as geographic concentrations of interconnected companies, specialized suppliers, service providers, and associated institutions in a particular field that are present in a nation or region.[68] Clusters arise because they increase the productivity with which companies can compete. They do this in three broad ways: first, by increasing the productivity of companies based in the area; second, by driving the direction and pace of innovation; and third, by stimulating the formation of new businesses within the cluster.[69] Achieving a critical mass with regard to the number of firms may be necessary for renewable electricity technology companies to develop the technological innovations enabling U.S. firms to be competitive in future markets. Competitive companies working with supplier industries and academic institutions in diverse geographic regions can aid the innovation cycle.

The close manufacturer-supplier relationships in clusters help companies to learn about technological needs or developments. Clusters make opportunities for innovation more visible, and provide the capacity and the flexibility to act rapidly. Local suppliers and partners can and do get closely involved in the innovation process, thus ensuring a better match with customer requirements.[70] Clusters competing with other similar regional clusters can increase the industry's competence domestically. Clusters are particularly well-suited to connections between manufacturing and technology, especially as these are advantaged by research institutions, and highlight the importance of a well-prepared workforce in enhancing productivity. If more green jobs is the goal, then these concepts may need to be applied to develop the renewable electricity industry. Cluster development could possibly trigger the type of innovation that can lead to the next generation of marketable renewable energy technologies resulting in opportunities for U.S. green jobs growth.

MARKET AND POLICY CHALLENGES FOR GREEN JOBS GROWTH

Renewable energy resources have an unmatched potential for clean energy production, if the right technologies can be employed to harness them. U.S. national energy policy with goals for use of renewable electricity and greenhouse gas mitigation will add to efforts in the states to grow the markets for renewable energy technologies. While the growth in power generation pointed to by EIA projections speaks to the potential market for sales of equipment and systems to provide renewable energy and hence manufacturing-related green jobs, the question of where the systems and components will be designed and built is quite another issue.

International Competition

The attraction of growing markets for clean energy and technologies will likely mean increasing international competition for renewable energy technology product orders. Most of the developed economies and a number of developing countries have or are seeking to create their own renewable electricity technology manufacturing sectors, and U.S. companies are helping other countries to develop or increase these capabilities.[71] Additionally, slowing renewable energy markets in Europe have turned the attention of European companies to the U.S. market. In fact, with the growing requirements from individual state renewable energy standards (and a possible federal requirement) spurring the sales of equipment and many Gigawatt-hours of renewable electricity sales for years to come, new entrants into the market could come from a number of countries with renewable energy expertise.

Existing companies with prior experience serving renewable energy markets will likely have an advantage. European-based companies (or their subsidiaries) currently dominate the wind turbine business in the United States.[72] These companies have marketing and technical expertise honed from serving their home markets, and are setting up joint ventures with U.S. firms (which can benefit from the knowledge base of their overseas partner). In such a setting, a new entrant which is purely a U.S. enterprise likely has a competitive disadvantage.

Such limitations are amplified in instances where the entrant is effectively subsidized by a national government. For example, China is seeking to be a global leader in sales of renewable energy equipment, and in a bid to build market share, is allegedly selling solar panels in the United States at prices less than the cost of materials, assembly and shipping. [73] Chinese companies are embracing foreign direct investment as part of this strategy with plans to build U.S. assembly plants for solar panel sold in the United States to avoid protectionist legislation.[74]

Incentivizing Domestic Production

A key to maximizing U.S. green jobs growth from renewable energy is the domestic design and manufacturing of equipment and components. The more of these functions that take place in the United States, the greater the number of higher paying jobs that will likely be located here. Plants to assemble components built and designed elsewhere miss the opportunity for intellectual capacity development and learning that may result in advances in the technology. These advances may eventually result in competitive advantages throughout a supply chain or manufacturing cluster. A renewable energy industry capable of serving the export market may create many more jobs than an industry which only serves domestic needs since production would necessarily be at internationally competitive prices. This can be achieved all the easier if the domestic market has sufficient demand to bring renewables rapidly down the cost curve.[75]

Many factors and considerations may influence the decision of where to build a factory. For example, labor costs overseas can be a fraction of those in the United States.[76] If such overhead costs for manufacturing products can be reduced and reflected in the final price of a product, the resulting competitive advantage could result in increased sales. For other firms, having a technically competent, well-educated workforce may be a paramount consideration. This may ease the training process for workers with an unfamiliar product and help accelerate the process of getting plant operations online and functioning.

Given the growing international competition for renewable energy markets and green jobs, policy mechanisms and incentives may be necessary to encourage manufacturers to locate production of renewable energy products and components in the United States. Incentives are commonly used by countries (and states within countries) to give a competitive advantage to domestic manufacturers. Some of the more frequently used policy tools include financial and tax incentives, local content requirements, and quality certification requirements. Each of these policy tools can be designed to favor domestic manufacturers in varying degrees. The degree to which such incentives can be used depend to a certain extent on the size of the domestic market, a factor which may favor U.S. production.[77]

Technological Challenges, Opportunities, and Risks

Renewable electricity technologies face well-defined challenges in making the transition from niche to mainstream power generation choices. Given the raw potential of certain renewable energy resources and results from ongoing research, further R&D investment is expected to increase the efficiency and applicability of the technologies.[78] However, research efforts do not always prove successful, and competitors may be quicker to bring improvements to the market. Incremental improvements may benefit a mature product, but most renewable energy technologies are not at that stage. Real breakthroughs in efficiency and applicability would benefit especially solar and wind energy technologies. Such innovations may provide opportunities to differentiate products and lead to competitive advantages, if these are adequately financed and appropriately focused on customer and market needs.

Developing the technical manufacturing capabilities needed to build today's and future generations of renewable electricity technologies may require better educated, highly trained workers capable of operating sophisticated equipment and controlling sensitive processes. A lack of qualified workers was identified as a potential barrier to growth of the U.S. renewable energy industry and was addressed by the Energy Independence and Security Act of 2007, Title 10, with additional funds subsequently provided by ARRA for green workforce training.[79] The opportunity exists for governments and educational institutions to work together with industry to assure that workers are properly prepared in sufficient numbers to meet anticipated increases in demand for electricity production from renewable resources.

Green Jobs as a Sustained National Focus

Over the last few decades, government interest and funding of renewable energy R&D has trailed the peaks and valleys of fossil energy prices. Most of the U.S. energy research budget has been spent on fossil fuels and nuclear energy, the old mainstays of central station generation.[80] But whatever advances in fossil fuels R&D may bring, legislative proposals to mitigate climate change concerns propose to levy a price on carbon emissions which could add significant costs to any future use of fossil fuels.

Few doubt the potential of renewable energy to help address climate change concerns; the question is whether the desired benefits merit the investment. While green jobs also encompasses jobs in energy efficiency and powering vehicles, growth of green jobs is generally envisioned as coming from the future development and deployment of renewable energy technologies. Renewable energy technologies will likely be an increasing part of the U.S. energy future. But developing the next generation of renewable energy technologies and building an internationally competitive industry may require a significant and sustained national investment. Without it, the majority of the solar panels, wind turbines, and components providing the clean energy of tomorrow may continue to be designed and built by workers overseas.

End Notes

[1] Nancy Spring, *CO_2: Climate Change Legislation* , Electric Light & Power, March 2008, http://uaelp.pennnet.com/ display_article/324458/34/ARTCL/none/none/1/C02:-Climate-Change-Legislation-/.

[2] The Utilities sector comprises establishments engaged in the provision of the following utility services: electric power, natural gas, steam supply, water supply, and sewage removal. U.S. Department of Labor, Bureau of Labor Statistics, *Industries at a Glance, NAICS 22*, July 2009, http://www.bls.gov/iag/tgs/iag22.htm.

[3] Energy Information Administration, *Primary Energy Production by Source*, Monthly Energy Review, August 2009, p. 5, http://www.eia.doe.gov/emeu/mer/pdf/pages/sec1_5.pdf.

[4] Most discussions today associate green jobs with renewable energy (i.e., the manufacture of wind turbines and solar panels, or building wind farms and solar parks) and energy efficiency (i.e., retrofitting existing homes and buildings with new energy-efficient windows, or constructing energy-efficient and environmentally-friendly office space).

[5] *Obama Administration's Plan for Energy: An Overview*, The White House, http://www.whitehouse.gov/issues energy_and_environment/.

[6] The United Nations World Commission on Environment and Development's 1987 report, *Our Common Future,* defines sustainability as "meeting the needs of the present without compromising the ability of future generations to meet their own needs."

[7] Sustainable Economies Consulting, *Green Jobs in North Carolina - 2008 Assessment*, June 2008, http://sustainableeconomies.com/nc_green_jobs_2008.pdf.

[8] An example of jobs resulting directly from the use of a renewable energy source, namely, wind, are the assemblers, installers, and maintainers of wind turbines. An example of jobs resulting indirectly from green economic activities are miners who excavate ore, mill workers who turn it into steel and metal workers who shape the steel into components of a wind turbine. An example of induced jobs are cashiers at grocery stores and ticket takers at movie theatres where workers in direct and indirect jobs spend their earnings.

[9] Ten full-time equivalent employees would be added to BLS to help develop statistics on green jobs that build upon the Bureau's Quarterly Census of Employment and Wages as well as its Occupational Employment Statistics and Employment Projections programs. The BLS anticipates producing, beginning in FY201 1, quarterly employment for establishments it determines engage in one or more green economic activities. In addition, statistics will become available on the concentration and characteristics of green occupations by industry.

[10] Management Information Services, Inc., *Defining, Estimating, and Forecasting the Renewable Energy and Energy Efficiency Industries in the U.S. and in Colorado*, The American Solar Energy Society, December 2008, pp. 6-7.

[11] Ibid., p. 8.

[12] Ibid., p. 26.

[13] Ibid., p. 36.

[14] Global Insight, *Current and Potential Green Jobs in the U.S. Economy*, U.S. Conference of Mayors, October 2008, p. 5.

[15] Although nuclear power generation is included in the 2006 count of green jobs, it is not part of the projection scenario.

[16] The studies are *The Work that Goes into Renewable Energy*, by the Renewable Energy Policy Project, November 2001; *Geothermal Industry Employment: Survey Results and Analysis*, Geothermal Energy Association, September 2005; and *California Renewable Technology Market and Benefits Assessment*, Electric Power Research Institute, November 2001.

[17] John M. Urbanchuk, *Contribution of the Ethanol Industry to the Economy of the United States*, Renewable Fuels Association, February 2009, http://www.ethanolrfa.org/objects /documents/2187/ 2008_ethanol_economic_contribution.pdf.

[18] Global Insight forecast one indirect job for every two direct jobs for both renewable electricity generation and renewable transportation fuels.

[19] Global Insight, *Current and Potential Green Jobs in the U.S. Economy*, U.S. Conference of Mayors, October 2008, p. 17.

[20] George Will, "Tilting at Green Windmills," *The Washington Post*, June 25, 2009.

[21] Gabriel Calzada, *Study of the Effects on Employment of Public Aid to Renewable Energy Sources*, King Juan Carlos University, March 2009, p. 27.

[22] The authors estimate, based on data from the Department of Energy's Energy Information Administration (EIA), that for retailers to supply from renewable sources 20% of total electricity in the coming decade an investment of $30 billion per year might be required.

[23] The analysts base their estimate that end-users will need to invest $3 billion annually over a 10-year period to increase their generation of electricity from renewable sources are based on EIA data.

[24] The researchers estimate that, if the market for ethanol from all sources is 20 billion gallons per year (*Energy Outlook 2009*), about $5 billion per year in additional investments would be required to "produce one-third of the quantity of ethanol from cellulosic sources by 2020." Robert Pollin, James Heintz, and Heidi Garrett-Peltier, *The Economic Benefits of Investing in Clean Energy*, Political Economy Research Institute, June 2009, p. 19.

[25] As noted in the introduction to this section, separate output and employment data are not available for electricity produced by wind and some other renewable resources (e.g., solar). In order to use an input-output model of the economy to estimate job creation, the authors developed definitions of these economic activities from industries uniquely identified in the North American Industry Classification System. For example, the authors defined wind-powered electricity as 26% construction industry; 12% plastic products, 12% fabricated metal, 37% machinery, 3% mechanical power transmission equipment and 3% electronic components manufacturing; and 7% scientific and technical services.

[26] Although fossil fuel electricity generation is uniquely identified in the North American Industry Classification, the authors note that oil-and-gas-powered energy production includes extractive, research, manufacturing and distribution activities. They accordingly defined oil-and-gas-powered energy as being 23% oil and gas extraction, 7% drilling oil and

gas wells, 4% support activities for oil and gas extraction, 10% natural gas distribution, 45% petroleum refineries, 8% petroleum product manufacturing, and 3% pipeline transport.

[27] Data users have ready access without confidentiality restrictions to NETS, unlike data of the U.S. Bureau of Labor Statistics and other government agencies. NETS is maintained by Walls & Associates, which licenses it to researchers.

[28] The Pew Charitable Trusts, *The Clean Energy Economy: Repowering Jobs, Businesses and Investments Across America*, June 2009, p. 10.

[29] Ibid., p. 12.

[30] Daniel Sosland, Derek Murrow, and Samuel Krasnow, *Energy Efficiency as Economic Stimulus*, Progressive Policy Institute, December 2008, http://www.ppionline. org/ndol/print. cfm?contentid=254849.

[31] State of Texas, Comptroller of Public Accounts, *Window on State Government: Overview of Renewable Energy*, http://www.cpa.state.tx.us/specialrpt/energy/renewable/index.php.

[32] Smart grid initiatives envision an advanced electricity generation, transmission, and distribution system which is capable of self-diagnosing and self-healing; incorporates numerous advanced technologies at all stages of the electricity supply chain; and focuses on efficiency, affordability and environmental quality. Massoud Amin, S. and Wollenberg, B.F., *Toward a Smart Grid: Power Delivery for the 21st Century*, Power and Energy Magazine, September 2006, http://www.todaysengineer.org/2008/Aug/smart_grid.asp.

[33] Changes in electric usage by end-use customers from their normal consumption patterns in response to changes in the price of electricity over time, or to incentive payments designed to induce lower electricity use at times of high wholesale market prices or when system reliability is jeopardized. Federal Energy Regulatory Commission, *A National Assessment of Demand Response Potential*, http://www.ferc.gov/ industries/ electric/indus-act/demand-response/drpotential.asp.

[34] Economists and other social scientists have demonstrated that the R&D activities of private firms generate widespread benefits enjoyed by consumers and society at large. As a result, the overall economic value to society often exceeds the economic benefits enjoyed by innovating firms as a result of their research efforts. Adam Jaffe, *Economic Analysis Of Research Spillovers Implications For The Advanced Technology Program*, Brandeis University and the National Bureau of Economic Research, December 1996, http://www.atp.nist.gov/eao/gcr708.htm.

[35] Dan Kammen, Kamal Kapdia, and Matthias Fripp, *Putting Renewables to Work: How Many Jobs can the Clean Energy Industry Generate?*, University of California, Berkeley, January 2006, (RAEL 2006) http://rael.berkeley.edu/ files/2004/Kammen-Renewable-Jobs-2004.pdf?src=.

[36] U.S. Department of Energy, Energy Information Administration (EIA), http://www.eia.doe.gov/glossary/ glossary_r.htm.

[37] Renewable energy resources include solar energy, wind, biomass, geothermal energy, and hydropower captured from the movement or temperature differentials of water.

[38] Ibid. p. 29.

[39] U.S. Department of Energy, Energy Information Administration (AEO 2009), *Annual Energy Outlook 2009*, Table 14 on p. 52, http://www.eia.doe.gov/oiaf/aeo/pdf/0383(2009).pdf.

[40] The present value of the total cost of building and operating a facility over its life, as represented by equal annualized amounts.

[41] U.S. Department of Energy, Energy Information Administration (AEO 2009), *Annual Energy Outlook 2009*, Table D10 on p. 186, http://www.eia.doe.gov/oiaf/aeo/pdf/0383(2009).pdf.

[42] EIA, *International Energy Outlook 2009*, http://www.eia.doe.gov/oiaf/ieo/highlights.html.

[43] AEO 2009.

[44] R.L. Bain, W.P. Amos, and M. Downing, et al., *Highlights of Biopower Technical Assessment: State of the Industry and the Technology*, National Renewable Energy Laboratory, NREL/TP-510-33502, April 2003, http://www.chemicalvision2020.org/pdfs/biopower _techassessment.pdf.

[45] AEO 2009.

[46] American Wind Energy Association, *WIND ENERGY GROWS BY RECORD 8,300 MW IN 2008*, http://www.awea.org/newsroom/releases/wind_energy_growth2008_27Jan09.html.

[47] CRS Report RL34546, *Wind Power in the United States: Technology, Economic, and Policy Issues*, by Stan Mark Kaplan.

[48] International Energy Agency, *Variability of Wind Power and Other Renewables: Management Options and Strategies*, Management Options and Strategies, 2005, http://www.iea.org/Textbase/Papers/2005/ variability.pdf.

[49] U.S. Department of Energy, Office of Energy Efficiency and Renewable Energy, *20% Wind Power by 2030: Increasing Wind Energy's Contribution to U.S. Electricity Supply*, December 2008, http://www1.eere.energy.gov/ windandhydro/pdfs/42864.pdf.

[50] Solar Today, *Game Changing Technology on the Horizon*, American Solar Energy Society, March 2009, http://www.solartoday-digital.org/solartoday/200903/?pg=27.

[51] Desert Protective Council, *Educational Bulletin #08-3*, http://www.dpcinc.org/_new/images

[52] U.S. Department of Energy, *Renewable Energy Annual 2007*, April 2009, http://www.eia.doe.gov/cneaf/ solar.renewables/page/rea_data/rea.pdf. (REA 2007).

[53] AEO 2009.

[54] Australian Broadcasting Corporation, *Germany an unlikely hot spot for solar power*, August 2007, http://www.abc.net.au/news/stories/2007/08/01/1994041.htm.

[55] Stanley R. Bull, *Renewable Energy: Status and Potential*, National Renewable Energy Laboratory, January 2009, (NREL 2009). http://www.drcog.org/documents/StanleyBull Presentation.pdf.

[56] NREL 2009.

[57] REA 2007.

[58] CRS Report R4063 1, *Water Issues of Concentrating Solar Power (CSP) Electricity in the U.S. Southwest*, by Nicole T. Carter and Richard J. Campbell.

[59] AEO 2009.

[60] Energy Information Administration, *Geothermal Heat Pump Manufacturing Activities, 2007*, February 2009, http://www.eia.doe.gov/cneaf/solar.renewables/page/ghpsurvey/ghps survey.html.

[61] Oak Ridge National Laboratory, *Dams: Multiple Uses and Types*, http://www.ornl.gov/info /ornlreview/rev26-34/ text/hydside1.html.

[62] Chris MacDonald, *Historic first: Emerging countries surpass developed nations in energy consumption*, June 2009, http://www.sustainable-development.com/economy/

[63] Organization for Economic Cooperation and Development.

[64] Demand-Side Management refers to the planning, implementation, and monitoring of utility activities designed to encourage consumers to modify patterns of electricity usage, including the timing and level of electricity demand. It refers to only energy and load-shape modifying activities that are undertaken in response to utility-administered programs. It does not refer to energy and load-shaped changes arising from the normal operation of the marketplace or from government-mandated energy-efficiency standards. Demand-Side Management covers the complete range of load-shape objectives, including strategic conservation and load management, as well as strategic load growth. http://www.eia.doe.gov/glossary/glossary_d.htm.

[65] Delegates agreed on principles for financing a fund to help the poorest nations cope with the effects of climate change at the Conference of the Parties meeting of the *United Nations Framework Convention on Climate Change* in Poznan, Poland, in December 2008.

[66] H.R. 2454, Subtitle D, Sec. 441.

[67] Wharton School, University of Pennyslvania, *Rising Giants: Industrial Clusters Are Changing the Face of Chinese Manufacturing*, June 2009, http://knowledge.wharton.upenn.edu/ article.cfm?articleid=2250.

[68] Harvard Business School, Institute for Strategy and Competitiveness, *Clusters and Cluster Development*, http://www.isc.hbs.edu/econ-clusters

[69] Michael E. Porter, *Clusters and the New Economics of Competition*, Harvard Business Review, December 1998, http://www.isc.hbs.edu/econ-clusters

[70] Ibid.

[71] Tom Friedman, *Have a Nice Day*, New York Times, September 15, 2009, http://www.nytimes.com/2009/09/16/ opinion/16friedman.html?_r=1&emc=eta1.

[72] Ryan Wiser and Mark Bolinger, *2008 Wind Technologies Market Report*, U.S. Department of Energy, July 2009, http://www1.eere.energy.gov/windandhydro/pdfs/46026.pdf.

[73] Keith Bradsher, "China Racing Ahead of U.S. in the Drive to Go Solar," *New York Times*, August 24, 2009.http://www.nytimes.com/2009/08/25/business/energy-environment/ 25solar.html?_r=1&em.

[74] Ibid.

[75] RAEL 2006.

[76] "Chinese wages are still less than $1 an hour, and factory workers in Vietnam earn as little as $50 a month for a 48- hour workweek, including Saturdays." Keith Bradsher, *Investors Seek Asian Options to Costly China*, New York Times, June 2008, http://www.nytimes.com/2008/06/18/business/worldbusiness/18invest.html?_r=1.

[77] Joanna Lewis and Ryan Wiser, *Fostering a Renewable Energy Technology Industry: An International Comparison of Wind Industry Policy Support Mechanisms*, Lawrence Berkeley National Laboratory, LBNL-591 16, November 2005, http://eetd.lbl.gov/ea/EMS/reports/59116.pdf.

[78] NREL 2009.

[79] CRS Report R40412, *Energy Provisions in the American Recovery and Reinvestment Act of 2009 (P.L. 111-5)*, coordinated by Fred Sissine.

[80] J. J. Dooley, *U.S. Federal Investments in Energy R&D: 1961-2008*, U.S. Department of Energy, October 2008, http://www.pnl.gov/main/publications/external/technical_ reports/PNNL-17952.pdf.

In: Federal Green Jobs Efforts
Editor: Alecia Gretter
ISBN: 978-1-63117-269-4

Chapter 2

EMPLOYMENT AND TRAINING: LABOR'S GREEN JOBS EFFORTS HIGHLIGHT CHALLENGES OF TARGETED TRAINING PROGRAMS FOR EMERGING INDUSTRIES*

United States Government Accountability Office

WHY GAO DID THIS STUDY

Labor received $500 million from the Recovery Act to help create, better understand, and provide training for jobs within the energy efficiency and renewable energy industries, commonly referred to as "green jobs." Since 2009, Labor has also "greened" existing programs and funded additional green jobs training grants and other efforts.

In light of the amount of funding targeted to green programs within Labor, GAO examined: (1) what is known about the objectives and coordination of Labor's green jobs efforts, (2) what type of green jobs training grantees provided and how selected grantees aligned their training to meet employers' green jobs needs, (3) what is known about program outcomes and what challenges, if any, grantees faced in implementing their programs, and (4) what Labor has done to assist and monitor its green jobs grantees. To conduct

* This is an edited, reformatted and augmented version of the United States Government Accountability Office publication, No. GAO-13-555, dated June 2013.

this work, GAO reviewed relevant federal laws and regulations; surveyed selected offices within Labor using two questionnaires—one for directly-funded green jobs efforts and one for other efforts; interviewed Labor officials and 11 out of 103 green jobs training grantees; and analyzed relevant Labor documents and data.

What GAO Recommends

GAO recommends that Labor identify lessons learned from the green jobs training programs to enhance its ability to implement such programs in emerging industries. Labor agreed with our recommendation.

What GAO Found

Of the $595 million identified by Labor as having been appropriated or allocated specifically for green jobs activities since 2009, approximately $501 million went toward efforts with training and support services as their primary objective, with much of that funding provided by the American Recovery and Reinvestment Act of 2009 (Recovery Act). Because the Recovery Act directed federal agencies to spend funds quickly and prudently, Labor implemented a number of high-investment green jobs efforts simultaneously. As a result, in some cases, Recovery Act training programs were initiated prior to a full assessment of the demand for green jobs, which presented challenges for grantees. While Labor's internal agencies initially communicated with each other and with other federal agencies after the Recovery Act was passed, most Recovery Act grants have ended or are winding down.

Labor created its green jobs definitional framework to provide local flexibility, and grantees we interviewed broadly interpreted Labor's framework to include any job that could be linked, directly or indirectly, to a beneficial environmental outcome. Labor's training data show most participants were trained in construction or manufacturing. While the findings of our site visits are not generalizable, all grantees we interviewed said they had worked closely with local employers to align their training program with the green skills needs of local employers. Most grantees we interviewed also told us they had incorporated green elements into existing training programs aimed at traditional skills, such as teaching weatherization as part of a carpentry training program.

The outcomes of Labor's green jobs training programs remain uncertain, in part because data on final outcomes were not yet available for about 40 percent of grantees, as of the end of 2012. Analysis of grantees with final outcome data shows they collectively reported training slightly more individuals than they had projected, but job placements were at 55 percent of the target. Training-related job placement rates remain unknown because Labor's Office of Inspector General (OIG) found these data unreliable. Grantees we interviewed were generally positive about Labor's green job training programs, but most said they had faced challenges during implementation, including: (1) a lack of reliable green jobs labor market information, (2) insufficient time to meet grant requirements, (3) knowledge gaps surrounding green skills and changing energy policies, and (4) difficulty placing participants into green jobs, primarily due to the overall poor economy.

Labor has provided technical assistance and taken steps to monitor green jobs training grantees through on-site monitoring visits and quarterly reviews. During these visits and reviews, Labor officials assessed grantee performance, such as by comparing reported program outcomes, including job placements, to targeted performance levels. However, Labor provided only limited guidance on how to document reported job placements. Labor officials required grantees with lower than projected performance levels to implement corrective action plans. In addition, Labor officials told us they have taken steps to improve the quality of grantee reported data, such as by forming an internal workgroup to identify ways to improve the technical assistance they provide to grantees on reporting performance outcomes.

ABBREVIATIONS

BLS	Bureau of Labor Statistics
CFBNP	Center for Faith-Based and Neighborhood Partnerships
CoP	Community of Practice
EISA	Energy Independence and Security Act of 2007
Energy	Department of Energy
ETA	Employment and Training Administration

ETP	Energy Training Partnership Grants
GCBG	Green Capacity Building Grants
GEMS	Grant Electronic Management System
GGS	Green Goods and Services Survey
GJIF	Green Jobs Innovation Fund
GTP	Green Technologies and Practices Survey
HUD	Department of Housing and Urban Development
HVAC Systems	Heating, Ventilation, and Air Conditioning Systems
ILAB	Bureau of International Labor Affairs
Labor	Department of Labor
MOU	Memorandums of Understanding
MSHA	Mine Safety and Health Administration
OA	Office of Apprenticeship
OASP	Office of the Assistant Secretary for Policy
ODEP	Office of Disability Employment Policy
OFCCP	Office of Federal Contract Compliance Programs
OIG	Officer of the Inspector General
O*NET	Occupational Information Network
OPA	Office of Public Affairs
OSHA	Occupational Safety and Health Administration
OWCP	Office of Workers' Compensation Programs
POP	Pathways Out of Poverty Grants
Recovery Act	American Recovery and Reinvestment Act of 2009
SESP	State Energy Sector Partnership and Training Grants
SLMII	State Labor Market Information Improvement Grants
TA Partnership	Technical Assistance Partnership
UI	Unemployment Insurance
VETS	Veterans' Employment and Training Services
WB	Women's Bureau
WHD	Wage and Hour Division
WIA	Workforce Investment Act of 1998

June 19, 2013

The Honorable Lamar Alexander
Ranking Member
Committee on Health, Education, Labor, and Pensions
United States Senate

The Honorable Michael Enzi
Ranking Member
Subcommittee on Children and Families
Committee on Health, Education, Labor, and Pensions
United States Senate

The American Recovery and Reinvestment Act of 2009 (Recovery Act) provided the Department of Labor (Labor) $500 million in funding aimed at training and placing workers into jobs in the "energy efficiency and renewable energy" industries,[1] commonly referred to as "green jobs." While many of these grants have ended or are currently winding down, they were initiated as part of a strategy aimed at moving the United States toward greater energy independence and security, and green jobs research and training programs were created that were expected to help American workers develop the specialized skills they would need to find and maintain employment in green jobs. Congress passed the framework for this investment—the Green Jobs Act of 2007[2]—more than 5 years ago, and according to Labor officials, funds for green jobs programs were not appropriated until the passage of the Recovery Act in 2009.[3] More recently, Labor has launched a number of efforts to "green" several of its existing programs, and has funded additional green jobs training grants, including the $38 million Green Jobs Innovation Fund (GJIF) grant program. In total, Labor funded 103 green jobs training grantees through the three Recovery Act-funded green jobs training programs and the GJIF program. In light of the significant amount of funding targeted to green jobs training programs within Labor, this review examines:

1) What is known about the objectives and coordination of Labor's green jobs efforts?
2) What type of green jobs training did grantees provide and how did selected grantees align their training to meet employers' green jobs needs?
3) What is known about program outcomes and what challenges, if any, did grantees face in implementing their programs?
4) What has Labor done to assist and monitor its green jobs grantees?

In addition, the Recovery Act requires that GAO conduct bimonthly reviews of how the act's funds are used by recipients.[4] As part of this review, we examined the use of Recovery Act funds along with other federal funding.

To address the first research question on the objectives and coordination of Labor's green jobs efforts, we compiled an inventory of Labor's green jobs efforts by requesting information from 14 different offices across Labor. These 14 offices were selected based on the likelihood of their administering a green jobs effort or program. 5 We asked these offices to list two separate sets of efforts: (1) efforts where federal funds were appropriated or allocated specifically for green jobs activities and, (2) efforts where federal funds were not specifically appropriated or allocated for green jobs activities, but where the office sought to incorporate green elements into either an existing program or ongoing activity. We gathered additional information on these efforts through two follow-up questionnaires and received fully completed questionnaires for nearly all efforts.[6]

Because the majority of Recovery Act funding for green jobs efforts was directed toward training programs, we focused much of our review on four grant programs—the three green jobs training programs funded by the Recovery Act (Energy Training Partnership grants, Pathways out of Poverty grants, and State Energy Sector Partnership and Training grants) as well as the newer Green Jobs Innovation Fund. To describe the characteristics of the 103 green jobs training grantees of these four grant programs, we also collected data from the department on the characteristics of green jobs grantees, including grantee location and organizational type.

To better understand the type of green jobs training grantees provided, how grantees aligned their training to meet green jobs needs, and what challenges, if any, they faced in implementing their programs, we interviewed 11 out of the 103 green jobs training grantees and analyzed data provided by Labor. We interviewed grantees in states that had a relatively high number of Labor green jobs grant recipients, states where grantees or sub-grantees received GJIF funds, and states in different geographic regions.[7] During our grantee interviews, we collected information about the types of green jobs training that were funded by Labor's green jobs training grants and the outcomes of grantees' programs, including green job placement. We specifically asked grantees about any challenges they may have encountered as they developed and implemented their program, including whether they experienced challenges placing participants into green jobs. We cannot generalize our findings beyond the interviews we conducted.

To assess the reliability of Labor's data on training type and outcomes, we (1) reviewed existing documentation related to the data sources, including Labor's Office of the Inspector General (OIG) reports, (2) electronically tested the data to identify obvious problems with completeness or accuracy, and (3) interviewed knowledgeable agency officials about the data. We determined that the data were sufficiently reliable for limited purposes. Specifically, we determined that training type data were sufficiently reliable for purposes of reporting out on the types of training most frequently provided by grantees. However, based upon the OIG's findings, we determined that the outcome data were not sufficiently reliable to determine the success of the programs and that data on the extent to which grantees entered training-related employment were not reliable enough to report.

To describe Labor's technical assistance and monitoring efforts, we reviewed technical assistance guides and Labor's Core Monitoring guide, interviewed Labor officials, and discussed Labor's technical assistance with selected grantees. We also obtained and reviewed copies of Labor's monitoring reports for green jobs training grantees. In conducting this review, we also reviewed relevant federal laws, regulations, guidance and pertinent Labor reports and procedures. A more detailed explanation of our scope and methodology can be found in appendix I.

We conducted this performance audit from May 2012 to June 2013 in accordance with generally accepted government auditing standards. Those standards require that we plan and perform the audit to obtain sufficient, appropriate evidence to provide a reasonable basis for our findings and conclusions based on our audit objectives. We believe that the evidence obtained provides a reasonable basis for our findings and conclusions based on our audit objectives.

BACKGROUND

The Workforce Investment System and Green Jobs Act of 2007

The Department of Labor oversees a number of employment and training programs administered by state and local workforce boards and one-stop career centers established under the Workforce Investment Act of 1998 (WIA). The green jobs training programs Labor has overseen were created under the Green Jobs Act of 2007, which amended WIA. The Green Jobs Act of 2007 was passed as part of the Energy Independence and Security Act of 2007 (EISA),[8] which was intended to move the United States toward greater energy independence and security and to increase the production of clean renewable fuels, among other objectives. This act directed the Secretary of

Labor to work in consultation with the Secretary of Energy to create a new worker training program to prepare workers for careers in the energy efficiency and renewable energy industries. However, funds for these programs were not appropriated until the passage of the Recovery Act in 2009, according to Labor officials.

Green Jobs and the Recovery Act

The Recovery Act appropriated $500 million in funding for competitive green jobs grant programs at Labor. The current administration presented the green jobs training grant program as part of a broad national strategy both to create new jobs and to reform how Americans create and consume energy. Specifically, the administration articulated a vision for federal investments in renewable energy to involve coordination across a number of federal agencies to create new, well-paying jobs for Americans and to make such jobs available to all workers.

Table 1. Green Jobs Grant Programs Funded by the Recovery Act

Program	Description	Eligible grantees	Training focused
Energy Training Partnership (ETP) Grants	Provided funds so grantees could develop partnerships and provide training and supportive services, such as registered apprenticeships, that would lead to employment for individuals in need of updated training and unemployed workers, among others	Private nonprofit organizations that are either national labor-management organizations with local networks or nonprofit partnerships—either statewide or local	Yes
Pathways Out of Poverty (POP) Grants	Provided funds to deliver training and placement services to individuals seeking pathways out of poverty, including unemployed individuals, high school dropouts, individuals with a criminal record, and disadvantaged individuals living in areas of high poverty by providing employment within the energy efficiency and renewable energy sectors	Private nonprofit organizations that deliver services through networks of local affiliates, for example, or whose local affiliates have the ability to provide services in four or more states or public or private local organizations such as community colleges or faith-based organizations	Yes

Program	Description	Eligible grantees	Training focused
State Energy Sector Partnership and Training (SESP) Grants	Provided funds for training, job placement, and related activities that reflect a comprehensive energy sector strategy aligned with the governor's overall workforce vision	State workforce investment boards, in partnership with local officials, employer and industry leaders, and labor organizations	Yes
State Labor Market Information Improvement (SLMII) Grants	Provided funds to collect, analyze, and disseminate labor market information and to enhance the labor exchange infrastructure for careers within the energy efficiency and renewable energy sectors	State workforce agencies or consortia of the state workforce agencies of multiple states	No
Green Capacity Building grants (GCBG)	Provided funds to build the capacity of existing Labor grantees to prepare targeted populations for employment in green industries and to support organizations so that they could bolster the capacity of their training programs through the purchase of equipment, staff professional development, curriculum development and/or adaptation, partnership development, and where necessary, the hiring of additional staff	Grantees that had already received a grant from Labor under specified grant programs, such as YouthBuild or Women in Apprenticeship and Non-Traditional Occupations	No

Source: GAO analysis of Labor's Solicitations for Grant Applications.

The Employment and Training Administration (ETA) was responsible for overseeing the implementation of the green jobs training programs that were authorized in the Green Jobs Act of 2007 and funded through the Recovery Act. In June 2009, ETA announced a series of five Recovery Act grant competitions related to green jobs, three of which were primarily focused on training. All of these programs are scheduled to end before the end of July 2013. Table 1 describes these five programs and identifies the types of organizations eligible to receive each grant.

Between September 2010 and October 2012, Labor's OIG issued a series of three reports related to the department's Recovery Act green jobs programs, including training programs.[9] The most recent report raised questions about the low job placement and retention of trained program participants, the short amount of time for which many participants received training, and limitations of available employment and retention data, among other things.

Labor's Broad Framework for Green Jobs

Labor has used a broad framework to define green jobs, incorporating various elements that have emerged over time as the understanding of what constitutes a green job has evolved. As part of the Green Jobs Act of 2007, WIA was amended to identify seven energy efficiency and renewable energy industries targeted for green jobs training funds.[10] In addition, beginning in 2009, Labor issued information on 12 emerging green sectors as part of a broader effort to describe how the green economy was redefining traditional jobs and the skills required to carry out those jobs.[11] Most recently, in 2010, the Bureau of Labor Statistics (BLS) released a two-part definition of green jobs that was used to count the number of jobs that could be considered green either because what the work produced or how the work was performed benefitted the environment.[12] Figure 1 shows the various elements Labor has used over time to define green jobs.

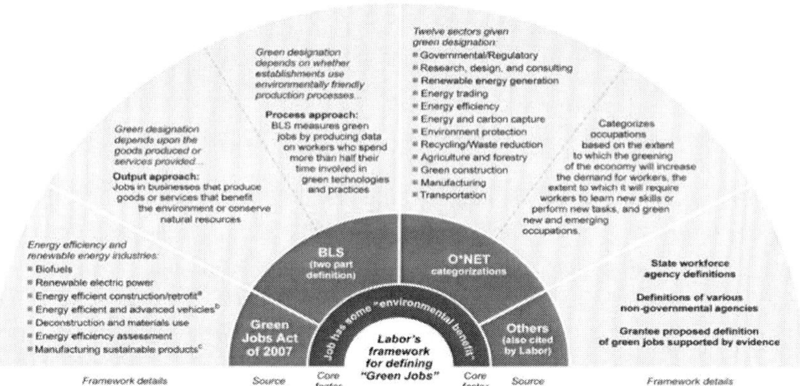

Source: Green Jobs Act of 2007 and GAO review of Labor documents.

[a] Energy efficient building, construction, and retrofit.

[b] Energy efficient and advanced drive train vehicle.

[c] Manufacturers that produce sustainable products using environmentally sustainable processes and materials.

Figure 1. Key Elements of Labor's Green Framework for Defining Green Jobs.

During the implementation of its Recovery Act green jobs training programs and the newer Green Jobs Innovation Fund (GJIF) pilot and demonstration project, which was initiated in 2011 to support job training opportunities for workers in green industry sectors and occupations, Labor has allowed program applicants to use combinations of the various elements to propose the type of training they would provide. For example, Labor permitted grantees of the three Recovery Act-funded programs to develop their green jobs training programs by focusing on training in the seven industries specified by the Green Jobs Act of 2007, or to propose strategies for training in sectors or occupations identified as green within the Occupational Information Network (O*NET). In addition to these two elements, Labor allowed applicants of the more recent GJIF program to use the BLS definition. Furthermore, Labor allowed GJIF applicants to propose their own definition of green jobs if they were able to support their definition with evidence.

LABOR HAS FOCUSED ITS GREEN JOBS EFFORTS LARGELY ON JOB TRAINING

Most Funding for Green Jobs Efforts Has Been Directed toward Training Workers

According to funding information provided by Labor and our survey of Labor's directly-funded green jobs efforts, most funding for green jobs efforts at Labor has been directed toward programs designed to train individuals for green jobs, with less funding supporting efforts with other objectives, such as data collection or information materials.[13] Indeed, approximately $501 million (84 percent) of the $595 million identified by offices at Labor as having been appropriated or allocated specifically for green jobs activities since 2009 went toward efforts with training and support services as their primary objective.[14] In total, approximately $73 million, or 12 percent of the total amount of funding for green jobs activities, was reported appropriated or allocated for data collection and reporting efforts.

Most of the funding for green jobs efforts was provided through the Recovery Act, which funded both training and non-training-focused projects at Labor in part to increase energy efficiency and the use of renewable energy sources nationwide. In addition to Recovery Act funding for green jobs efforts, funding information provided by Labor and through our survey of directly-funded green jobs efforts indicate that Labor has allocated at least an additional $89 million since 2009 to support seven other green jobs efforts that have been implemented by five of Labor's offices (see fig. 2). For a brief description of each of Labor's green jobs efforts for which funds were appropriated or allocated, see appendix II.

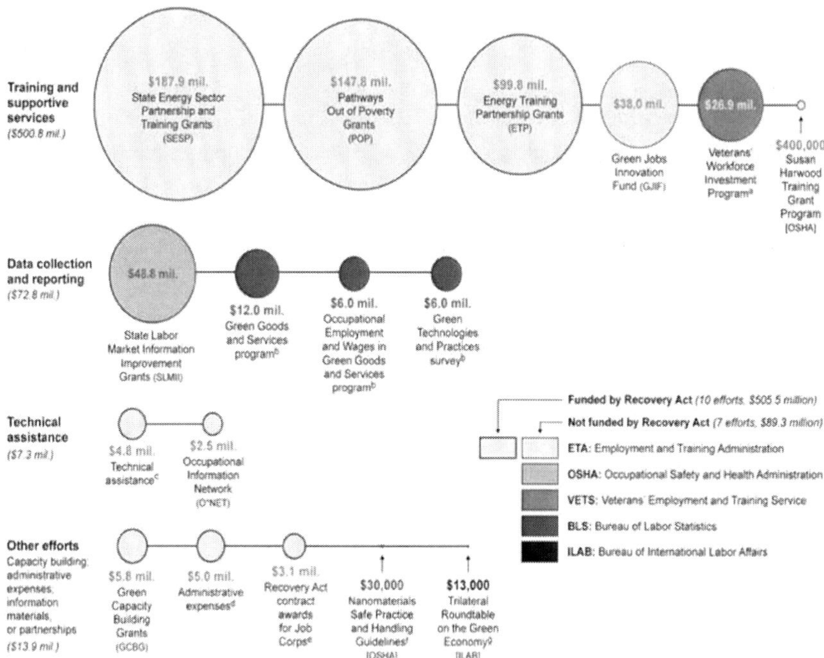

Source: GAO analysis of survey data for directly-funded data provided by Labor, and additional funding identified by GAO.

[a] This figure represents cumulative funding for the program for fiscal years 2009, 2010, and 2011. The amount appropriated or allocated for the program was $7,641,000, $9,641,000, and $9,622,000, respectively.

[b] Since its inception in fiscal year (FY) 2010, through FY 2012, BLS has been appropriated approximately $24 million for the Green Jobs initiative. Of this amount, BLS allocated about $4 million annually for the Quarterly Census of Employment and Wages portion of the Green Goods and Services (GGS) survey; about $2 million annually for the Occupational Employment Statistics portion and; about $2 million annually for the Green Technologies and Practices (GTP) survey. The Employment Projections program was also funded from the same appropriation specifically for green jobs activities.

[c] This figure represents approximately $3.4 million in Recovery Act funding for technical assistance provided through the National Governors Association and its partners for green jobs grantees and approximately $1.5 million in non-Recovery Act funding for technical assistance for GJIF grantees.

Funding for technical assistance for Health Care Sector and Other High Growth and Emerging Industries Grantees is excluded from this funding figure.

[d] Administrative expenses include funding for three green jobs evaluations: A Labor Market Information evaluation ($499,980), a Green Jobs and Health Care Implementation Report ($917,248), and a 5-year impact evaluation ($7,992,852).

[e] Recovery Act contract awards for Job Corps primarily supported capacity building activities, though some funding was used for training activities.

[f] Recovery Act funds were used to fund this effort.

[g] Trilateral Roundtable: The Employment Dimension of the Transition to a Green Economy (February 3-4, 2011); this figure represents an expenditure instead of an appropriated or allocated amount.

Figure 2. Funding for Green Jobs Efforts at Labor since 2009, by Office and Focus Area.

The Recovery Act directed federal agencies to spend the funds it made available quickly and prudently, and Labor implemented a number of relatively brief but high-investment green jobs efforts simultaneously. As a result, in some cases, Recovery Act training programs were initiated prior to a full assessment of the demand for green jobs.[15] Specifically, Recovery Act-funded green jobs training grantees designed and began to implement their green jobs training programs at the same time states were developing green job definitions and beginning to collect workforce and labor market information on the prevalence and likely growth of green jobs through the State Labor Market Information Improvement grants, which were also funded with Recovery Act funds. Furthermore, BLS launched its Green Jobs initiative—which included various surveys designed to help define and measure the prevalence of green jobs—after many green jobs training programs had begun.[16] ETA officials noted that BLS's development of the definition of green jobs was a deliberative and extensive process that required consulting stakeholders and the public. They also said that BLS's timeline for defining green jobs differed from ETA's timeline for awarding and executing grants, which was driven by Recovery Act mandates.

Labor has made subsequent investments that build upon lessons learned through the Recovery Act grant programs. For example, ETA initiated the $38 million GJIF program in 2011 to support job training opportunities for workers in green industry sectors and occupations. In developing the GJIF grant program, ETA considered lessons learned through the Recovery Act grant programs.[17] For example, various stakeholders including employers, the public workforce system, federal agencies, and foundations identified Registered Apprenticeship—training that combines job related technical instruction with structured on-the-job learning experiences for skilled trades and allows participants to earn wages—as a valuable workforce strategy. ETA acknowledged that upgrading basic skills, including literacy and math, is critical to ensure job placement and suggested that training participants exclusively in green skills is not always sufficient. Consequently, ETA required GJIF grantees to implement green jobs training programs that would either forge linkages between Registered Apprenticeship and pre-apprenticeship programs or deliver integrated basic skills and occupational training through community-based organizations. Figure 3 shows a timeline illustrating the rollout of selected green jobs grants at Labor and the time periods during which these grants were active,[18] as well as the timing of BLS's efforts to collect data on green jobs. With the exception of the GJIF grants, all of these efforts will have been completed by July 2013.

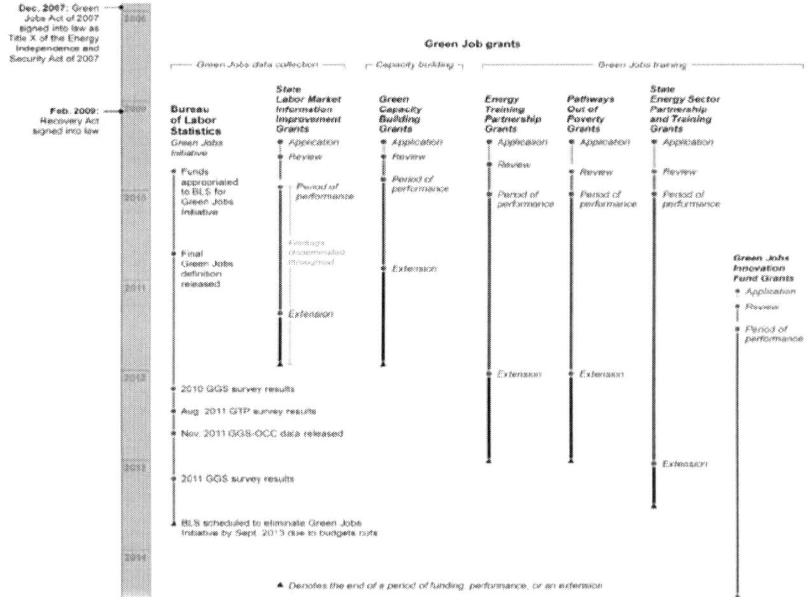

Source: GAO analysis of Labor documents.

Figure 3. Timeline of Selected Green Jobs Efforts at Labor.

Distribution and Characteristics of Green Jobs Training Grantees

Grantees from all but six states received at least one of the 103 green jobs training grants that were awarded by ETA,[19],[20] but grantees were somewhat concentrated within certain regions of the country. Specifically, most states with three or more grantees were located in the Northeast, West, or Midwest regions of the country.[21] Four states and the District of Columbia received five or more green jobs training grants: California, Michigan, New York, and Pennsylvania. Figure 4 shows the number of green jobs training grants awarded by state.

Source: GAO analysis of Labor data.

Figure 4. Green Jobs Training Grantees, by State.

In terms of organizational type, most green jobs training grants were awarded to nonprofit organizations and state workforce agencies (see fig. 5). Specifically, 44 percent of green jobs training grants were awarded to nonprofit organizations and 34 percent were awarded to state governmental agencies or departments. In addition, 10 percent of grantees were organized labor or labor management organizations. ETA officials from all six of its regional offices said that in terms of organizational type, green jobs training grantees did not differ substantially from the types of grantees ETA typically oversees. ETA officials said building partnerships had been an important focus of the green jobs grants, and indeed ETA's grant solicitations required, or in some cases encouraged, grant recipients, regardless of organizational type, to develop partnerships with various stakeholders, such as representatives of the workforce system, industry groups, employers, unions, the education and training community, nonprofits, or community-based organizations. Staff from ETA's regional offices said that some grantees developed new and successful partnerships as a result of the grants, including partnerships with labor unions.

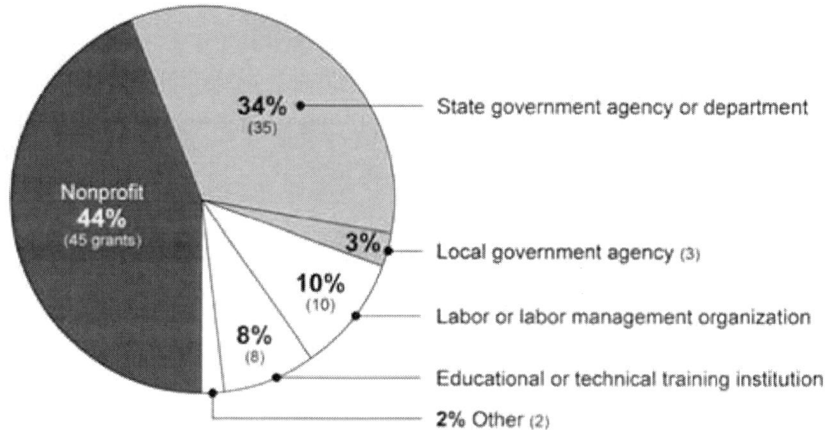

Source: GAO analysis of Labor data.

Figure 5. Green Jobs Training Grantees, by Organizational Type.

More than half of ETA's green jobs training grantees implemented their grants through sub-grantees, or a network of local affiliates, rather than providing training services directly to participants. Grantees that contract with sub-grantees or local affiliates to provide services are responsible for monitoring and overseeing how all grant funds are used, effectively delegating day-to-day oversight responsibility from Labor to the primary grantee.[22]

Labor's Efforts to Infuse Green Jobs into Existing Programs

In addition to Labor's direct investments in green jobs, several offices at Labor have infused green elements into their ongoing activities even though funds were not specifically appropriated or allocated for these green jobs efforts. In total, of the 14 Labor offices we surveyed, 6 identified and implemented 48 such efforts (for a list of the efforts, see appendix III). Some of these offices added a "layer of green" to existing training programs or other activities. For example, according to material provided by Labor, most

YouthBuild programs have incorporated green building into their construction training. Other efforts focused on providing information materials, forming partnerships, or conducting publicity and outreach, among other things. For example, the Women's Bureau[23] created a guide on sustainable careers for women and Labor's Occupational Safety and Health Administration contributed to an Environmental Protection Agency publication on best practices for improving indoor air quality during home energy upgrades. Further, in 2010 the Center for Faith-Based and Neighborhood Partnerships hosted a roundtable discussion about green jobs between the Secretary of Labor and leaders from national foundations and discussed how to create employment opportunities for low-income populations in the green jobs industry.

Several of Labor's Offices Coordinate on Implementing Green Jobs Efforts within Labor or with Other Federal Agencies

Although funding for green jobs efforts at Labor has shifted and green jobs efforts funded through the Recovery Act are winding down, a few of Labor's ongoing programs or efforts continue to emphasize green jobs or skills, and Labor continues to incorporate green elements into existing programs by coordinating internally on an as-needed basis. After the passage of the Recovery Act, a number of Labor's offices worked together to implement the requirements of the act, and Labor officials said that they collaborated on green jobs efforts on a fairly regular basis and that more formal green jobs meetings across the department were common. For those green jobs efforts where green elements have been infused into ongoing activities even though funds were not specifically appropriated or allocated for green jobs efforts, offices at Labor indicated through our survey that they continue to coordinate on such efforts within Labor and across other federal agencies, albeit in a less formal manner. For example, according to our survey of these indirectly-funded green jobs efforts, for 37 of 46 of the efforts listed in appendix III, offices said that they coordinated with others at Labor, and for 30 of 46 of the efforts, they reported coordinating with other federal agencies.[24] In addition, it is likely that coordination on green jobs efforts will continue to occur on an ad-hoc basis, especially as funding and priorities within the department shift. For example, Labor recently reported that due to federal budget cuts, BLS has discontinued its reporting on employment in green jobs.

According to a Labor official, after the Recovery Act was passed, Labor collaborated with other departments, such as the Department of Energy (Energy) and the Department of Housing and Urban Development (HUD) to foster job growth for a new green economy. For example, Labor's Occupational Safety and Health Administration worked with Energy on retrofitting and safety activities, and Labor also partnered with HUD to provide green jobs training and possible employment opportunities to public housing residents. In addition, Labor entered into various Memorandums of Understanding (MOU) after the Recovery Act was passed to collaborate on green jobs-related issues with other federal agencies. For example, the Secretaries of Energy, Labor, and the Department of Education announced a collaboration to connect jobs to training programs and career pathways and to make cross-agency communication a priority. While these examples highlight coordination on green jobs efforts after the passage of the Recovery Act, little is known about the effectiveness of these efforts.

Selected Grantees Broadly Defined Green Jobs and Generally Added Green Elements to Existing Training Programs

To identify the potential demand for green jobs in their communities, all (11 of 11)[25] grantees we interviewed had broadly interpreted Labor's green jobs definitional framework to include as green any job that could be linked, directly or indirectly, to a beneficial environmental outcome. While Labor created its framework to provide local flexibility, the wide variation in the types of green jobs obtained by program participants illustrates just how broadly Labor's definition can be interpreted and raises questions about what constitutes a green job—especially in cases where the job essentially takes the form of a more traditional job (see table 2). In general, grantees we interviewed considered jobs green if they could link the job to (1) a green industry, (2) the production or installation of goods that benefit the environment, (3) the performance of services that potentially lead to environmental benefits, or (4) environmentally beneficial work processes. For example, in some cases, grantees we interviewed considered jobs green because they were linked to the renewable energy industry, such as solar panel installation or sales. Grantees considered other jobs green because the goods being produced benefited the environment, such as the pouring of concrete for a wind turbine or the installation of energy efficient appliances. In some cases the green job was service-based, such as an energy auditor or energy surveyor. Finally, other grantees considered jobs green because of the environmentally beneficial processes being used, such as applying paint in an efficient manner or using advanced manufacturing techniques that reduce waste. Even for jobs where parts of the work have a link to environmentally beneficial outcomes, workers may only use green skills or practices for a portion of the time they work. For instance, technicians trained to install and repair high-efficiency heating, ventilation, and air conditioning (HVAC) systems may in the course of their work also install less energy efficient equipment.[26]

Table 2. Examples of Green Job Placement Based on Interviews with Selected Labor Green Jobs Grantees

Training focus	Green job placement examples
Renewable energy	• solar panel installer • solar panel sales representative
Green construction and retrofitting	• construction laborer building an energy efficient home • laborer inspecting and weatherizing homes • operator driving heavy machinery on a public rail project • concrete pourer on a wind-turbine project site • plumber installing low-flow toilets and water saving fixtures • hardwood floor layer installing sustainable flooring materials • worker trained to apply paint in a manner that reduces waste and toxic emissions
Electrical work	• electrician installing energy-efficient lighting controls • maintenance advisor on energy-saving appliances and lighting alternativesfor property managers
Heating, Ventilation, and Air Conditioning (HVAC) Systems	• HVAC technician installing high-efficiency furnaces and air conditioning units • worker who conducts energy surveys to educate building owners on cost-effective energy efficiency alternatives • maintenance technician overseeing compliance on EPA emission regulations
General and hazardous waste management	• laborer on an environmental clean-up project • laborer removing hazardous waste from a factory site
Manufacturing	• factory line-worker trained to identify opportunities to reduce waste • machinist using computerized technology that reduces waste
Other	• emergency medical technician trained to perform rescues on wind turbine towers

Source: GAO analysis of information from selected Labor green jobs training grantees.

All grantees we interviewed said they had worked closely with local employers to align their training program with the green skills needs of local employers. All agreed developing effective relationships with employers was crucial to aligning any training program with available jobs.[27] Labor's three Recovery Act green jobs training programs, as well as the GJIF program, all required applicants to demonstrate how they would partner with local employers to develop and implement their training programs. Most (9) grantees told us they had assembled advisory boards consisting of representatives from local businesses and industry associations to help inform them about available green jobs and the skills that would most likely be in demand by local employers. Further, all grantees said they engaged in ongoing communication with employers to stay abreast of changes in the local economy and employer needs, and most (10) made changes to their program curricula or tailored their training in response to employer input.

Labor's data show that green jobs training grantees primarily offered training in the construction and manufacturing industries. Specifically, nearly half of all participants of the Recovery Act-funded green jobs training programs received training focused on construction, and approximately 15 percent received training in manufacturing. Over 5 percent of participants in those programs received training in other industries that included utilities, transportation, and warehousing. Grantees in Labor's newer GJIF program focused even more heavily on construction—approximately 94 percent of participants were trained in construction and around 3 percent in manufacturing.

Most grantees (9) we spoke to had infused green elements into existing training curricula for more traditional skills. However, the extent to which the training focused on green versus traditional skills varied across programs and often depended upon the skill level of targeted participants. Most (7) of the programs we visited generally targeted relatively low-skilled individuals with limited work experience and were designed to teach participants the foundational skills they would need to pursue a career in a skilled trade in which green skills and materials can be used. For example, those programs typically used their green job grant funds to incorporate green skills into existing construction, carpentry, heating/airconditioning, plumbing, or electricity programs. The programs generally involved a mixture of classroom and hands-on training and taught traditional skills, such as how to read blueprints, use tools, install and service appliances, and frame buildings. In teaching these skills, however, instructors also showed students the way the processes or products used in performing these tasks could lead to

percent of participants who had exited between April 1, 2011, and March 31, 2012, had entered employment.[35] However, the UI data do not capture whether jobs obtained were training-related for either the Recovery Act-funded or GJIF programs, so, absent additional relevant information, the extent to which grantees placed participants into training-related employment may never be reliably known.

According to Labor officials, once complete, these additional UI wage data may provide more definitive information on the extent to which program participants entered employment and will be used by the department to develop a broader picture of the grant programs' level of success in achieving employment outcomes. Specifically, Labor officials said that while there is not a formal process to study the UI data, program staff routinely examine these data to identify lessons learned and best practices that could be applied to future grant programs. Labor officials said the data could be used to compare the green jobs training programs against other training programs across the agency, such as those under WIA, if resources permit. While Labor officials consider the UI data to be more definitive than the grantee-reported job placement data to measure overall program outcomes once the grant period ends, they stressed the importance of having real-time data to monitor grantee performance during implementation.

While the UI wage record data provide an alternative source of information on job placement outcomes, due to a 9-month lag time, these data are of limited usefulness regarding program management. Specifically, because of the time lag, grantees could not use these data to monitor their progress toward meeting program goals in real-time. Further, Labor could not use the data to hold grantees accountable for meeting grant goals, as all grant periods will have ended before the data are complete. Consequently, ensuring the reliability of grantee reported outcome data remains vitally important, particularly for grant programs whose primary objective is to prepare workers for attaining employment in a targeted emerging industry.

Grantees Highlighted Green Jobs Training Benefits and Challenges

The grantees we interviewed were generally positive about Labor's green jobs training programs, with most speaking optimistically about the potential value of the green skills obtained by the program participants. Most grantees we met with said that they believe there to be a continued national movement towards lowering energy usage—whether due to economic, policy, or cultural changes[36]—and all projected that the demand for workers with green skills credentials will continue to rise. All (11 of 11) were of the opinion that possessing green skills in addition to more traditional skills provides workers with an advantage as they seek a new job or move along a career pathway, and most (10) cited the need for training programs that provide nationally or industry-recognized green credentials. Two noted that having multiple credentials was particularly valuable. Lastly, some (5) grantees mentioned that the benefits of the green jobs training, like most job training, may not become apparent immediately, but may often be realized later during the worker's career, especially as demand for green skills grows.

However, all grantees noted there have been challenges associated with developing and implementing Labor's green jobs training programs. For example, most (8) of the grantees we interviewed said that the lack of credible green jobs labor market information had limited their ability to identify or predict the level of available green jobs or the demand for green skills in their local area. Although state workforce agencies received funding to conduct green jobs labor market information studies under the State Labor Market Information Improvement grants, most resulting data were issued after many Recovery Act training programs had already begun. In addition, the BLS surveys were released from March 2012 through March 2013, after GJIF grantees had submitted their applications outlining their training programs to Labor. Having access to the final results of the state labor market information studies could have provided Recovery Act grantees with additional insights into their state's economic activity in the energy efficiency and renewable energy industries, as well as jobs within those industries when they were developing their training programs. Similarly, BLS survey results could have provided GJIF grantees with a national snapshot of establishments that produce green goods and services and the jobs of workers involved in green activities, among other information, and may have provided grantees with additional context for the development and implementation of their green jobs training programs.[37] Labor officials said the rapidly evolving nature of the green industries has resulted in multiple changes to employer green job demand information over the course of the grant periods, further complicating their attempts to provide labor market information for this sector.

In addition, most (9) grantees we met with said Labor's green jobs training grants did not afford them enough time to both develop local partnerships and recruit, train, and place program participants. All grantees said developing partnerships can be especially time-consuming if such partnerships had not existed prior to the grant award. 38 Most (9) noted that given how important local partnerships are to developing successful training programs, training programs that require such partnerships should have longer grant periods than those afforded by the Recovery Act and GJIF programs.

Furthermore, most (8) grantees mentioned how developing and implementing a relatively new type of training, like green skills, can require additional time in order to fill knowledge gaps among employers. This may be especially true in light of changing state and local energy policies. For example, according to the Department of Energy, as of March 2013, 29 states have established standards aimed at generating a certain percentage of the state's energy using renewable sources by a specified year. Furthermore, many municipalities throughout the country are requiring that local construction projects adhere to environmentally friendly requirements. However, most (9) grantees we spoke with said some employers may not recognize how changing policies will affect their businesses. In fact, they believe this lack of understanding may be limiting demand for workers trained in green skills. To address this problem, one of the grantees we interviewed had developed a 1-day training program for local business managers to educate them about how they could benefit from the green skills that participants were obtaining through the organization's training program. Most (6) grantees said at times during the implementation of their green jobs training program, they were, in effect, attempting to simultaneously drive both supply and demand for workers with green skills, which took considerable time and effort.

In addition, although all grantees we interviewed had engaged with employers who had committed their support for the training curriculum, they also said this did not always translate into green jobs for program participants. Above all, most (9) pointed to the slow economic recovery as the reason their predictions—and those made by employers—regarding green job growth were not fully realized. For example, one grantee explained how the local housing market had not recovered as quickly as anticipated, and as a result, demand for workers with green skills—such as green construction techniques, weatherization practices, and the installation of energy efficient appliances—has been sluggish. In addition, most (10) grantees explained that because green skills are often intertwined with traditional skills training and the skilled labor industries, their programs' participants were negatively affected by the overall poor economy. For example, most grantees (9) noted how their program participants, despite their additional layer of green skills training, found themselves competing with a high number of unemployed workers who were also seeking to regain employment in more traditional jobs such as carpentry or electrical work. Most (9) grantees also noted that renewable energy sectors, such as solar power, have not grown in their regions as was predicted several years ago. Lastly, most (7) grantees we interviewed said it is difficult to accurately measure the value of green skills training in terms of green job placement.

In general, they said this is partly because, unlike jobs in other growing industries, like health care, there are few distinctly green jobs. One grantee we met with said she believes the term "green job" is misleading, and complicates program implementation. This grant official said that funding should be directed toward supplementing traditional skills training with green skills that can be used on any job rather than on preparing workers for specific jobs identified as green.

Based on our interviews with grantees of Labor's green jobs training programs, and the descriptions of their experiences implementing those programs, we identified several lessons learned that may warrant consideration when implementing similar targeted grant programs for other emerging industries (see table 3).

Table 3. Lessons Learned Associated with Implementing Grant Programs for an Emerging Industry

Implementing grant programs for an emerging industry may require…	Lessons learned based on selected grantees' experiences implementingLabor's green jobs training programs
…a more nimble approach	• Setting realistic outcome goals can be complicated by uncertainty regarding the economy and public policies. • If available labor market information is limited, significant, ongoing employer involvement will likely be required. • Local market conditions may quickly change necessitating modifications to training and grant agreements. • There can be unanticipated challenges to training and placing target populations into jobs as the industry continues to evolve.
… longer grant periods	• Building effective relationships with employers and key partners can require a large investment of time and effort. • Identifying industry-recognized credentials may require significant research. • If the emerging industry is intertwined with the skilled labor trades, preparing low-skilled workers to be job-ready will likely require supplemental training and assistance.

Implementing grant programs for an emerging industry may require…	Lessons learned based on selected grantees' experiences implementingLabor's green jobs training programs
… a raised awareness of the emerging industry	• Learning about the new industry and staying current requires continuous research. • Consensus on the value of specific job credentials may be lacking. • Training providers and employers are not always aware about how changing policies will affect the demand for new worker skills.
… more effective measurement of program outcomes	• If the industry is defined broadly to allow local flexibility, it may limit the comparability of reported outcome data. • Accurate counting of outcomes is difficult when the definition of the industry continues to evolve. • Training for jobs of the future may result in delayed outcomes if jobs are slow to materialize.

Source: GAO analysis of interviews with selected grantees.

LABOR HAS ASSISTED AND MONITORED GRANTEES, BUT HAS PROVIDED LIMITED GUIDANCE ON OUTCOME DOCUMENTATION REQUIREMENTS

Labor Has Provided Various Forms of Technical Assistance to Support Green Jobs Grantees

Labor has provided all green jobs grantees with technical assistance to help them implement their grant programs and comply with relevant federal laws and regulations. For example, Labor officials have hosted technical assistance webinars on topics such as financial management and how to engage employers. Labor also maintains a website for each green jobs training grant program and a green jobs community of practice on its online platform, Workforce3One. In addition, Labor has published bimonthly digests for Recovery Act grantees since January 2011 that highlight new technical assistance materials and other grant-related information. Finally, Labor has compiled and periodically updated a technical assistance guide that briefly describes and provides hyperlinks for its technical assistance resources, including webinar recordings and promising practices. Several grantees we interviewed (4 of 11) reported participating in webinars and referring to technical assistance materials posted to Workforce3One.

In addition, ETA has funded three separate studies to assess the implementation of selected green jobs programs funded by the Recovery Act. Specifically, Labor funded a 2-year implementation evaluation that examined the implementation of the three Recovery Act-funded green jobs training programs and issued both interim and final reports.[39, 40]

Labor also funded an evaluation of the State Labor Market Information Improvement grants and issued a final report and additional related products in 2013.[41] Finally, Labor has funded an ongoing impact evaluation scheduled to be completed in 2016. This study was designed to test the extent to which selected grantees of one of the four green-jobs training programs overseen by ETA—Pathways Out of Poverty— improved worker outcomes by imparting skills and training valued in the labor market.[42]

To support its technical assistance efforts to grantees, Labor entered into a grant agreement with the National Governors Association, which together with two partner organizations formed a Technical Assistance Partnership (TA Partnership).[43] In conjunction with Labor officials, the TA Partnership has facilitated monthly conference calls for each grant program so grantees can learn from their peers and receive program-specific technical assistance. The TA Partnership has also compiled and updated reports that highlight promising practices grantees have implemented. Finally, the TA Partnership and Labor officials have held annual grantee conferences, which have covered various topics including strategies to retain and place program participants and the importance of nationally recognized credentials. Several (4 of 11) grantees we interviewed mentioned participating in the monthly conference calls and annual conferences and said that generally they had been helpful.

While Labor provided guidance and technical assistance on how to document eligibility for the green jobs training programs, it provided little guidance on what documentation grantees were expected to maintain regarding program outcomes, particularly with respect to job placement.

Specifically, while Labor provided guidance on how to report required performance data into its Recovery Act Database, this guidance does not specify what documentation, if any, grantees were to maintain for reported job placements, including those considered training-related. Our *Standards for Internal Control in the Federal Government* provides that internal control and all transactions and other significant events should be clearly documented and readily available for examination.[44] However, in its last green jobs report, the OIG found that nearly a quarter of reported outcomes were not supported by adequate documentation. One regional official noted that sub-grantees may not have known what documentation was required and staff in another office said that in some cases primary grantees may not have done enough to ensure that the sub-grantees they were responsible for overseeing understood documentation requirements.

While Labor officials have not issued additional guidance to GJIF grantees regarding how to document job placement and retention outcomes, they said they have taken other steps that address the OIG's recommendation to improve the quality of grantee reported performance data and utilize lessons learned from Recovery Act-funded green jobs training programs for other discretionary grant programs. First, ETA officials noted that they have formed an internal workgroup focused on improving the technical assistance provided to ETA's discretionary grantees about how to report program outcomes. This group hopes to issue recommendations in September 2013, and ETA officials believe these recommendations will help improve grant application instructions, and help ETA refine their reporting systems, among other things. Second, ETA officials told us that they had initiated a grant re-engineering project in August 2012 to identify common grant management challenges and develop strategies for addressing such challenges. For instance, the group has discussed ways to improve ETA's grant solicitation process, such as by including clearer expectations and benchmarks for performance in its solicitations for grant applications and by taking steps to ensure greater comparability of goals across grantees. Labor hopes to begin implementing the group's recommendations for new discretionary grant programs in August 2013.

Labor Monitored Grantees Using a Risk-Based Approach

ETA monitors most grants, including its green jobs training grants, through a risk-based strategy that prioritizes monitoring activities based upon grantees' assessed risk-levels and availability of resources, among other factors, and is described in its Core Monitoring Guide.[45] Specifically, according to officials from all six of ETA's regional offices, ETA's federal project officers monitor grantees as part of their ongoing duties,[46] which include calling grantees to offer technical assistance. In addition, ETA's federal project officers perform quarterly desk reviews, during which they review financial reports and quarterly performance reports that grantees are required to submit.[47] For the green jobs training grants, these reports include information such as the total amount of grant funds spent, the number of participants who began or completed training, a timeline for grant activities and deliverables, grantee accomplishments, and technical assistance needs. During these quarterly reviews, federal project officers compare grantees' reported performance outcomes and spending rates to those goals set by grantees in their grant proposals. Based upon their review of each grantee's reported information, federal project officers enter information about each grantee into Labor's Grant Electronic Management System (GEMS), which assesses risk and generates a risk level for each grantee.

The GEMS assessment of each grantee's risk level is then used by Labor to develop its risk-based monitoring strategy, which involves prioritizing site visits based on grantees' assessed risk-levels and availability of resources, among other factors. [48]According to regional officials from all six offices, nearly all green jobs training grantees received at least one on-site monitoring visit, typically about halfway through the period of performance. During these site visits, federal project officers assessed grantees' management and performance and documented any noncompliance findings and requirements for corrective action, as necessary. For example, Labor's site visit guide includes questions for federal project officers to consider about financial and performance data reporting systems and performance outcomes.

As a result of its on-site monitoring activities, Labor officials identified and required certain grantees to correct a variety of issues concerning the management of their grants. Many monitoring reports for the Recovery Act-funded green jobs training grants indicated that grantees were not on track to meet their performance outcomes. In such cases Labor required grantees to submit written corrective action plans that described what strategies they would undertake to increase project outcomes and how they would ensure that remaining funds would be used in a timely way to accomplish project objectives.

Labor officials said that grantees have made significant progress toward attaining their goals for beginning and completing training as a result of both the grantees' own efforts and ETA's technical assistance and monitoring efforts. These officials also stressed that while ETA holds grantees accountable to adhering to their grant statements of work, grantees are not contractually obligated to meet performance outcomes.[49] Unlike contracts or WIA-funded programs, which can impose sanctions for failing to meet projected targets, the accountability mechanisms for these green jobs grant programs were more limited. For example, ETA officials said that if a grantee does not achieve its placement outcomes, this can affect whether the grantee receives a period of performance extension for the current grant or, potentially, a future grant from ETA. Officials said that they had not withdrawn funding from any grantees for failing to meet performance targets for any of the four green jobs training programs. However, in some cases ETA officials decided not to grant extension requests for grantees reporting poor performance. As a result, some grant funds remained unexpended and will be returned to the Treasury, as required.

In addition to insufficient progress toward targeted outcomes, the monitoring reports of the Recovery Act-funded green jobs training grantees identified other noncompliance findings, including insufficient monitoring of sub-grantees. For example, a number of monitoring reports indicated that primary grantees had not sufficiently monitored their sub-grantees. These findings are notable given that such a large percentage of grantees implemented their programs through a network of sub-grantees. Both GAO and the Department of Justice's OIG have stressed the importance of sufficient sub-recipient monitoring to the grant oversight process.[50] Other noncompliance findings included grantees lacking adequate documentation to show program participants were eligible for services or grantees having failed to follow acceptable procurement processes. According to officials from all six regional offices, federal project officers did not identify any instances of fraud, waste, or abuse during their on-site monitoring visits.

CONCLUSION

The Recovery Act funded multiple, substantial investments in training programs targeted to a specific emerging industry—energy efficiency and renewable energy. Most of these programs have already ended or are currently winding down, although a few of Labor's continuing programs, such as YouthBuild, have incorporated many green elements since 2009, and the Green Job Innovation Fund program is scheduled to remain active through June of 2014. Despite the sizeable investment in green jobs, the green jobs training programs have faced a number of implementation challenges and final outcomes remain uncertain, particularly regarding placement into green jobs. A number of these challenges have stemmed from the need to implement the grants quickly and simultaneously before green jobs had been defined and more had been learned about the demand for green skills. Others, such as problems with the reliability of outcome data, can be traced to management issues that have compromised Labor's ability to measure the program's success, particularly regarding placing participants into training-related employment. Specifically, because Labor did not establish clear and timely guidelines for how to document green job placement outcomes, Labor is not able to assess the extent to which the targeted green jobs training programs placed participants in employment related to the training they received.

The challenges for an emerging industry such as energy efficiency and renewable energy are substantial. Uncertainty and debate still surround the question of what constitutes a green job. Under Labor's current framework, almost any job can be considered green if a link between the employee's tasks and environmental benefits can be made. Indeed, most grantee officials we interviewed said that most green jobs they have trained participants for are primarily traditional skilled-trades jobs, such as carpentry or electrical work. Many have been termed "green" because the worker has been trained to be mindful of energy use and reduce waste, or has been placed where the worker's tasks resulted in a product or service that benefited the environment, such as a light-rail construction site.

Such an approach provides certain benefits within the context of an emerging industry, in that many of the skills workers obtain can be transferred to traditional jobs in cases where local demand for green jobs falls below expectations. It also may serve to raise general worker awareness about energy efficiency and waste reduction, to the benefit of the employer or nation. Nonetheless, this emphasis on training that often takes the form of traditional skills training with an added layer of green may not fully align with the intent of the targeted training funds.

By funding several evaluations of green jobs training and labor market information programs, Labor has positioned itself to build upon lessons learned through implementing these individual programs. A fundamental consideration is whether it is prudent to implement job training programs for an emerging industry before more is known about the demand for skills and workers. Another consideration is whether it would be more or less effective for federally-funded training programs to focus on providing valuable green skills and credentials applicable on a wide variety of jobs, rather than to devote considerable attention to what is defined as a green job.

Even though Labor is scaling back its own green jobs efforts, energy efficiency and renewable energy will likely remain a national priority. Labor has established a green jobs community of practice on its online platform, Workforce3One, which, if maintained and used, can continue to facilitate information-sharing among grantees and workforce professionals regarding what green skills and credentials employers in their communities value most. In addition, the substantial investment in energy efficiency and renewable energy made through these grant programs also provides Labor an opportunity to identify broader lessons learned about the challenges and benefits associated with offering targeted training in an emerging industry, which could help inform the development of training for other emerging industries in the future. Without the benefit of such lessons learned and a continued focus on what is needed to address emerging industries, state and local workforce entities may grapple with similar challenges in the future.

RECOMMENDATIONS FOR EXECUTIVE ACTION

To enhance Labor's ability to implement training programs in emerging industries, GAO recommends that the Secretary of Labor identify lessons learned from implementing the green jobs training programs. This could include:

- Identifying challenges and promising strategies associated with training workers for emerging industries—through both targeted grant programs and existing programs—and considering ways to improve such efforts in the future. For example, taking a more measured or multi-phased approach could allow the time necessary to better determine demand for an emerging industry and establish the partnerships needed to properly align training with available jobs.
- Taking steps to ensure training programs adequately document outcome variables, particularly for targeted programs where tracking training relatedness is of particular interest.

AGENCY COMMENTS AND OUR EVALUATION

We provided a draft of this report to the Department of Labor. Labor agreed with our recommendation. Specifically, Labor's response noted that the department has already begun assessing lessons learned from the implementation of its green jobs grants. Labor also cited efforts to compile lessons learned to inform the design and implementation of future grant initiatives, including new approaches to capture program outcomes. Labor agreed that documenting outcomes is important and said it will work to provide technical assistance to ensure grantees adequately document outcomes. Finally, Labor noted the department will continue to collect information on employment outcomes and wages and will analyze these data once they are complete to provide a more definitive and final picture of the extent to which former green jobs training participants entered and retained employment.

Andrew Sherrill
Director
Education, Workforce, and Income Security Issues

APPENDIX I: OBJECTIVES, SCOPE, AND METHODOLOGY

Our objectives were to determine: (1) what is known about the objectives and coordination of the Department of Labor's (Labor) green jobs efforts, (2) what type of green jobs training grantees provided and how selected grantees aligned their training to meet employers' green jobs needs, (3) what is known about program outcomes and what challenges, if any, grantees faced in implementing their programs, and (4) what Labor has done to assist and monitor its green jobs grantees. To address these objectives, we reviewed relevant federal laws, regulations, and departmental guidance and procedures. We also created a data collection instrument and two questionnaires to obtain information from Labor officials. In addition, we analyzed data from Labor and interviewed selected grantees by phone or in person in five states—California, Illinois, Louisiana, Minnesota, and Pennsylvania—as well as Labor officials. We conducted this performance audit from May 2012 through June 2013 in accordance with generally accepted government auditing standards. Those standards require that we plan and perform the audit to obtain sufficient, appropriate evidence to provide a reasonable basis for our findings and conclusions based on our audit objectives. We believe that the evidence obtained provides a reasonable basis for our findings and conclusions based on our audit objectives.

Data Collection for Labor's Green Jobs Efforts

Our data collection strategy for obtaining information on green jobs efforts across Labor consisted of two phases.

First, we created a data collection instrument to obtain information on green jobs efforts across Labor. In the data collection instrument, we asked offices[1] at Labor to list two separate sets of efforts: (1) efforts where federal funds were appropriated or allocated specifically for green jobs activities and (2) efforts where federal funds were not specifically appropriated or allocated for green jobs activities, but where the office sought to incorporate green elements into either an existing program or ongoing activity. We distributed the data collection instrument to 14 of Labor's 28 offices: Occupational Safety and Health Administration (OSHA), Mine Safety and Health Administration (MSHA), Women's Bureau (WB), Employment and Training Administration (ETA), Veterans' Employment and Training Services (VETS), Office of the Assistant Secretary for Policy (OASP), Bureau of International Labor Affairs (ILAB), Bureau of Labor Statistics (BLS), Center for Faith-Based and Neighborhood Partnerships (CFBNP), Office of Federal Contract Compliance Programs (OFCCP), Wage and Hour Division (WHD), Office of Workers' Compensation Programs (OWCP), Office of Disability Employment Policy (ODEP), and the Office of Public Affairs (OPA). These 14 offices were selected based on the likelihood of their administering a green jobs effort or program. For example, we did not distribute the data collection instrument to Labor's Office of Inspector General, Office of the Solicitor, or Office of the Chief Financial Officer.

Second, we used the information we collected on the two separate sets of green jobs efforts in the data collection instruments to inform two follow-up questionnaires. For the first set of green jobs efforts, offices at Labor initially identified 16 efforts where funds were specifically appropriated or allocated for green-job related activities. For each of the 16 efforts, we sent a questionnaire by e-mail. The questionnaire focused on (1) the goals and objectives of the green jobs efforts, (2) how green jobs were defined for each of the efforts, (3) whether offices coordinated with others on these efforts, and (4) funding levels for each of the efforts. We pretested the questionnaire with two respondents from OSHA in December and made revisions. We then sent the questionnaires out on a rolling basis between January 16 and February 22, 2013. We determined 2 of the 16 efforts to be out of scope. Of the remaining 14 directly-funded green jobs efforts across five offices (OSHA, ETA, VETS, ILAB, and BLS), we received completed questionnaires for 13 and one partially completed questionnaire by April 3, 2013. [2] We also identified 3 additional directly-funded efforts, for a total of 17 efforts.

For the second set of green jobs efforts, offices at Labor initially identified 54 efforts where funds were not specifically appropriated or allocated for green jobs efforts, but green elements were incorporated into existing programs or ongoing activities. We identified two additional green efforts that fall under this category. We sent a brief questionnaire consisting of two questions by e-mail in an attached Microsoft Word form. The two questions included in the questionnaire were pre-tested as part of the more detailed survey mentioned above. All questionnaires were sent on January 29, 2013, or on February 22, 2013. We determined 10 of the 56 efforts to be out of scope. Of the remaining 46 efforts across six offices (OSHA, WB, ETA, VETS, ILAB, and CFBNP), we received completed questionnaires for all 46 efforts by March 21, 2013. Labor later identified 2 additional efforts, for a total of 48 efforts.

Because the majority of Recovery Act funding for green jobs efforts were directed toward training programs, we focused much of our review on four grant programs—the three training- focused green jobs training programs funded by the Recovery Act (Energy Training Partnership grants, Pathways Out of Poverty grants, and State Energy Sector Partnership and Training grants) as well as the newer Green Jobs Innovation Fund. To report on the characteristics of Labor's 103 green jobs training grantees, we obtained data from Labor on each training-focused green jobs grant administered by ETA. Specifically, we obtained information on the grantee's location, organizational type, and whether or not the grantee had sub-grantees.

Selected Grantees and Analysis of Program Outcome Data

To better understand the type of green jobs training grantees provided, how grantees aligned their training to meet green jobs needs, and what challenges, if any, they faced in implementing their programs, we analyzed data from Labor and interviewed 11 out of the 103 green jobs training grantees between August 2012 and April 2013. We conducted site visits in four states and interviewed grantees in two additional states by phone. We visited grantees in California, Illinois, Minnesota, and Pennsylvania, and interviewed grantees in Connecticut and Louisiana by phone. We selected grantees in these states because these states had a relatively high number of Labor green jobs grant recipients, grantees in these states received GJIF grants, and the states varied in their geographic locations. We selected both Recovery Act- and GJIF- funded green jobs training grantees, but emphasized GJIF-funded grantees since unlike many of the Recovery Act programs, the GJIF program is still active.

During each site visit we interviewed Labor's green jobs training grant officials, training providers, local employers, and, to the extent possible, program participants. Similarly, during our phone calls we interviewed grant officials and in one case employers. During the interviews, we collected information about the types of green jobs training that were funded by Labor's green jobs training grants and the outcomes of grantees' programs, including the impact of the training with respect to green job placement, or otherwise. We specifically asked grantees about any challenges they may have encountered as they developed and implemented their program, including whether they experienced challenges with respect to placing participants into green jobs. In addition, we collected information on how local employers were involved in the development of the training programs and the green job opportunities they were able to offer program participants. We cannot generalize our findings beyond the interviews we conducted.

To assess the reliability of Labor's training type and outcome data, we (1) reviewed existing documentation related to the data sources, including Labor's Office of Inspector General (OIG) reports, (2) electronically tested the data to identify obvious problems with completeness or accuracy, and (3) interviewed knowledgeable agency officials about the data. We determined that the data were sufficiently reliable for limited purposes. For example, we determined that training type data were sufficiently reliable for purposes of reporting out on the industries for which grantees most frequently trained participants. We included information about the extent to which Recovery Act-funded green jobs training grantees collectively reported meeting their enrollment, training completion, and entered employment targets for those grantees for which final data were available as of December 31, 2012. However, based upon the OIG's findings, we determined that the outcome data were not sufficiently reliable to determine the success of the programs. Finally, based upon the OIG's findings, we determined that the data on the extent to which grantees entered training-related employment were not reliable enough to report, even compared to targeted levels.

Analysis of Labor's Technical Assistance and Monitoring Efforts

To describe Labor's technical assistance efforts, we reviewed technical assistance guides and material posted to Workforce3One, interviewed Labor officials, and discussed Labor's technical assistance with selected grantees. To describe and assess Labor's monitoring efforts, we reviewed its Core

Monitoring guide, interviewed Labor officials in Washington, D.C. and in each of ETA's six regional offices—Atlanta, Boston, Chicago, Dallas, Philadelphia, and San Francisco—and obtained and reviewed copies of Labor's monitoring reports for green jobs training grantees, including recipients of Energy Training Partnership, Pathways Out of Poverty, and State Energy Sector Partnership and Training grants.

APPENDIX II: LABOR'S GREEN JOBS EFFORTS SINCE 2009 FOR WHICH FUNDS WERE APPROPRIATED OR ALLOCATED, BY FOCUS AREA

Focus Area: Training and supportive services		
Office	Name of effort	Reported description
ETA	Energy Training Partnership (ETP) grants	Through the Energy Training Partnership Grants, ETA awarded nearly $100 million to 25 projects. Grantees were to provide training and placement services in the energy efficiency and renewable energy industries to workers impacted by national energy and environmental policy, individuals in need of updated training related to the energy efficiency and renewable energy industries, and unemployed workers. Grantees were required to partner with labor organizations, employers and workforce investment boards. Grant awards ranged from approximately $1.4 to $5 million.
ETA	Pathways Out of Poverty (POP) grants	In total, Pathways Out of Poverty grantees received approximately $150 million in Recovery Act funds. The grant aimed to help targeted populations find pathways out of poverty through employment in energy efficiency and renewable energy industries. Grants ranged from approximately $2 million to $8 million and were awarded to eight national nonprofit organizations with localaffiliates and to 30 local public organizations or private nonprofit organizations.
ETA	State Energy Sector Partnership and Training (SESP) grants	Through SESP, ETA awarded nearly $190 million to state workforce investment boards in partnership with state workforce agencies. The grants were designed to provide training, job placement, and related activities that reflect a comprehensive statewide energy sector strategy including the governor's overall workforce vision, state energy policies, and training

Focus Area: Training and supportive services		
Office	Name of effort	Reported description
		activities that lead to employment in targeted industry sectors. ETA made 34 awards that ranged from approximately $2 to $6 million each.
ETA	Green Jobs Innovation Fund (GJIF)	The Green Jobs Innovation Fund was authorized under the Workforce Investment Act to help workers receive job training in green industry sectors and occupations and access green career pathways. In total, $38 million in grant funds were awarded to six organizations with networks of local affiliates to develop green jobs training programs. These programs were required to incorporate green career pathways either by forging linkages between Registered Apprenticeship and pre-apprenticeship programs or by integrating the delivery of technical and basic skills training through community-based partnerships.
ETA	Job Corps	Job Corps is a residential job training program for at-risk youth. The Job Corps program aims to teach participants the skills they need to secure a meaningful job, continue their education, and be independent. Job Corps has instituted a number of measures in recent years to "green" its job training programs and facilities. Recovery Act funding was used to incorporate "green" training elements into the automotive, advanced manufacturing, and construction trades at Job Corps centers nationwide and to pilot three new "green" training programs at selected Job Corps centers: Solar Panel Installation, Weatherization, and SmartGrid technology.
OSHA	Susan Harwood Training Grant Program	Targeted topic training grant in which applicants propose training based on the occupational safety and health topics chosen by OSHA. Alternative Energy Industry Hazards and Green Jobs Industry Hazards were included as topics in FY 2009 and FY 2010, respectively.
VETS	Veterans' Workforce Investment Program (VWIP)	VWIP supports veterans' employment and training services to help eligible veterans reintegrate into meaningful employment and to stimulate the development of effective and targeted service delivery systems. In FYs 2009 and 2010, project proposals received priority

(Continued)

Focus Area: Training and supportive services		
Office	Name of effort	Reported description
		consideration if they supported "Green Energy Jobs" and proposed clear strategies for training and employment in the renewable energy economy.
Focus Area: Data collection and reporting		
Office	Name of effort	Reported description
ETA	State Labor Market Information Improvement (SLMII) grants	ETA awarded approximately $48.8 million in State Labor Market Information Improvement Grants to support the research and analysis of labor market data to assess economic activity in energy efficiency and renewable energy industries and identify occupations within those industries. Grant activities included collecting and disseminating labor market information, enhancing strategies to connect job seekers to green job banks, and helping ensure that workers find employment after completing training. ETA awarded 30 grants of between $763,000 and $4 million.
BLS	Green Goods and Services (GGS) program	This is a survey-based program, covering 120,000 business establishments, which provides a measure of national and state employment in industries that produce goods or provide services that benefit the environment.
BLS	Occupational Employment and Wages in Green Goods and Services (GGS-OCC) program	This program provides occupational employment and wage information for businesses that produce green goods and services.
BLS	Green Technologies and Practices (GTP) survey	This is a special survey of business establishments designed to collect data onestablishments' use of green technologies and practices and the occupations of workers who spend more than half of their time involved in green technologies and practices.
BLS	Employment Projections program	Green Career Information staff within the Employment Projections program produces career information on green jobs including wages, expected job prospects, what workers do on the job, working conditions, and necessary education, training, and credentials.

Focus Area: Technical assistance		
Office	Name of effort	Reported description
ETA	Occupational Information Network (O*NET)	Information that O*NET provided was used to support all Recovery Act green jobs grantees who were doing green jobs data collection and training in the states. These Recovery Act funds to O*NET were for the specific purpose of focusing occupational research and data collection on green jobs on an accelerated pace.
ETA	Technical Assistance	The Technical Assistance Partnership led by the National Governor's Association supported Recovery Act-funded green jobs grantees.
Focus Area: Other - Capacity building, administrative expenses, information materials, partnerships		
Office	Name of effort	Reported description
ETA	Green Capacity Building Grants (GCBG)	In total, ETA awarded $5 million in Recovery Act funds to training programs already funded by the Department of Labor to build their capacity to provide training in the energy efficiency and renewable energy industries. ETA awarded 62 of these grants, with awards ranging from $50,000 to $100,000.
ETA	Administrative Expenses	ETA used $5 million of the $500 million authorized for the Recovery Act green jobs grants for administrative expenses (salaries and expenses). This does notinclude any funds that were retained for technical assistance for these grants. Administrative expenses were in part used to fund three separate evaluations of Recovery Act green jobs programs: (1) a Labor Market Information evaluation, (2) a green jobs and health care implementation report, and (3) a 5-year impact evaluation.
OSHA	Nanomaterials Safe Practice and Handling Guidelines	This guidance was funded by the Recovery Act and is a guidance document for R&D workers and employers in the nanotechnology field.
ILAB	Trilateral Roundtable: The Employment Dimension of the Transition to a Green Economy (February 3-4, 2011)	The U.S. Department of Labor, Human Resources and Skills Development Canada and the European Commission brought together U.S., Canadian, and European experts representing governments, trade unions, industry, and nongovernmental organizations to discuss the transition to the green economy.

(Continued)

Focus Area: Other - Capacity building, administrative expenses, information materials, partnerships		
Office	Name of effort	Reported description
		Discussions focused on defining and measuring green jobs, establishing effective green jobs partnerships, designing green skills development and training, ensuring green jobs serve as a pathway out of poverty, and examining the quality of green jobs, as well as the sustainability of green jobs investments by governments.

Source: Responses provided by Labor.

APPENDIX III: LABOR'S GREEN JOBS EFFORTS SINCE 2009 FOR WHICH FUNDS WERE NOT APPROPRIATED OR ALLOCATED, BY FOCUS AREA

Focus Area: Information materials		
Office	Name of effort	Reported description
ETA	Office of Apprenticeship (OA)	In June 2009, Labor/ETA/OA published a report entitled, "The Greening of Registered Apprenticeship: An Environmental Scan of the Impact of Green Jobs on Registered Apprenticeship and Implications for Workforce Development." More recently, as part of the 75th Anniversary of the National Apprenticeship Act in 2012, OA put out a call to sponsors across the county to collect Registered Apprenticeship Innovators or Trailblazers. This process identified anumber of innovative programs across the country, including several specific examples of apprenticeship programs with a focus on green efforts.
ILAB	OECD Working Party on Employment; April 11, 2012	Labor officials participated in a technical review of economic research presented in "What Green Growth Means for Workers and Labour Market Policies: An Initial Assessment." Subsequently the paper appeared as Chapter4 in the 2012 OECD Employment Outlook.

Focus Area: Information materials		
Office	Name of effort	Reported description
OSHA	Green Job Hazards Web-site	An OSHA website providing green job safety information on specific green jobs, such as green roofing, waste management, wind energy, recycling, weatherization, and geothermal industries.
OSHA	Workforce Guidelines for Home Energy Upgrades	Provided safety information for weatherization jobs in collaboration with Department of Energy, Environmental Protection Agency, and National Institute for Occupational Safety and Health (NIOSH)
OSHA	EPA's Healthy Indoor Environment Protocols for Home Energy Upgrades	OSHA worked with EPA in the publication of this guidance which identifies critical indoor environmental quality risks and worker assessment protocols, and provides guidance to address these issues.
OSHA	Green and Clean: How Hospitals Can Protect Patients and Workers by Using Earth-Friendly and Sustainable Products/Practices	Through OSHA and The Joint Commission and Joint Commission Resources (JCR) Alliance, JCR developed an article that discusses the importance of adopting sustainable products and practices for cleaning, sanitizing, and disinfecting healthcare facilities. The article also provides requirements for selecting green cleaning products (January 2013).
WB	Why Green is Your Color, Manual	Publishing of a manual: Why Green Is Your Color: A Woman's Guide to a Sustainable Career. Designed to assist women with job training and career development.
WB	Teleconference Calls and Fact Sheets	A series of teleconferences for workforce practitioners about how to connect women with green jobs. A fact sheet accompanied each teleconference.
Focus Area: Partnerships		
Office	Name of effort	Reported description
CFBNP	Facilitated a partnership between OSHA regional office and green jobs grantee	In May 2010, the Deputy Director of CFBNP facilitated a partnership between OSHA's Cincinnati office and East End Community Services in Dayton, OH – aPathways Out of Poverty sub-grantee seeking a training module on safe handling of asbestos and lead removal as part of a green jobs training program.

(Continued)

Focus Area: Partnerships		
Office	Name of effort	Reported description
ETA	Interagency working groups (with Energy, Education, and HUD)	Interagency working groups related to clean energy and green jobs in which Labor works with the U.S. Departments of Education, Energy, and Housing and Urban Development
ILAB	March 21, 2012 Anniversary Paper from the 2011 Trilateral Roundtable on the Transition to a Green Economy	The U.S. Department of Labor, Human Resources and Skills Development Canada and the European Commission each contributed two updates on promising practices or strategies that were first explored at the Roundtable
ILAB	2012 G20 Labor and Employment Ministers' meeting, May 17-18, 2012	International labor and employment ministers met to discuss employment issues, including green jobs and high youth unemployment rates
ILAB	Participated in OECD's Employment, Labor and Social Affairs Committee meeting that discussed a green jobs project	Labor officials participated in the October 13-14, 2011 ELSAC meeting in Paris, France. One topic discussed at the meeting was the OECD's green jobs project.
ILAB	Rio+20 Conference on Sustainable Development input	Labor staff articulated labor and employment priorities to the U.S. interagency for inclusion in U.S. government positions for Rio+20, including for the U.S. position paper and during negotiations of the Rio outcome document.
ILAB	International Labor Organization Governing Body	Labor staff participated in International Labor Organization Governing Body discussions on green jobs and sustainable development
ILAB	Asia-Pacific Economic Cooperation Symposium on Human Capital Policies for Green Growth and Employment	The two-day Symposium convened experts from 16 Asia-Pacific Economic Cooperation member economies and international organizations to discuss sustainable economic development policies. The event was hosted by the Department of Education, in partnership with Labor.

Focus Area: Partnerships		
Office	Name of effort	Reported description
ILAB	US-Brazil Memorandum of Understanding on Labor Cooperation (March 20-21, 2012)	U.S. Secretary of Labor and her counterpart from Brazil signed a Memorandum of Understanding on Labor Cooperation in May 2012. The memorandum highlights cooperation in the area of green jobs.
ILAB	Information exchange with officials from Brazil's Ministry of Environmental Affairs about U.S. Labor's Green Jobs Initiatives	The Women's Bureau Director led a Labor delegation meeting with officials from Brazil's Ministry of Environmental Affairs at U.S. EPA about the definition of green jobs, and initiatives in both countries.
ILAB	Inter-American Conference of Ministers of Labor, of the Organization of American States	The October 2012 conference working group meetings considered green jobs in follow-up to the XVII IACML Declaration and Plan of Action adopted by the ministers of labor of the Americas in November 2011. The Plan of Action called for specific follow-up actions related to green jobs including, inter-alia, in-depth exchange of best practices in the region.
ILAB	African Growth and Opportunity Act's African Women Entrepreneur Program Briefing	DOL officials met with 47 women leaders from Sub-Saharan Africa under the African Growth and Opportunity Act's African Women Entrepreneur Program, sharing best practices, perspectives and strategies to train and employ women in green jobs.
OSHA	Federal Workgroup on Spray Polyurethane Foam	Federal workgroup (EPA, OSHA, NIOSH, CPSC) working with industry groups (ACC, CPI, SPFI) on environmental, occupational, and consumer health and safety issues with the use of spray polyurethane foam weatherization products
OSHA	OSHA/American Wind Energy Association Alliance	Presentations by Deputy Assistant Secretary of Labor for Occupational Safety and Health to general session, sharing of knowledge, development of informational products and participation in quarterly meetings.
OSHA	Recovery through Retrofit: Meeting	OSHA and Labor participated in an interagency Recovery through Retrofit Working Group comprised of over 80 technical staff members from the Departments of Energy, Housing and Urban Development, and Labor; Environmental Protection Agency; ,and USD that drafted standards for workers who will be involved in retrofitting homes to make them more energy

(Continued)

Focus Area: Partnerships		
Office	Name of effort	Reported description
		efficient. The group met in Denver for a 3 days and a follow-up meeting was held in Washington D.C. This is the Vice President's initiative. As a part of the working group, OSHA provided technical advice and input in the worker protection aspects of the standards that were drafted.
OSHA	National Alliance with the Professional Landcare Network (PLANET)	One of the goals of this Alliance is to participate in PLANET's Green Industry Conference
Focus Area: Publicity and outreach		
Office	Name of effort	Reported description
CFBNP	Secretary of Labor and CFBNP Hold Green Jobs Discussion with Philanthropic Leaders	On December 1, 2010, Secretary of Labor and Assistant Secretary for Employment and Training met with leaders from several national foundations to discuss significant investments in green jobs programs, as well as effective strategies that create employment and advancement opportunities for low-income populations in the green job industry.
CFBNP	Publicity of RecycleForce green jobs program for ex-offenders	In 2012, the Director of CFBNP wrote a blog for Fatherhood.gov about an Employment and Training Administration's grantee RecycleForce that providesgreen jobs to ex-offenders.
ETA	Job Train's "Earth Day Every Day Campaign"	Held the week of April 19th-23rd, 2010, the campaign was designed "to raise environmental awareness among students and staff and serve as friendly reminders to be more energy efficient."
ILAB	Briefing for Chinese Embassy official	Labor staff briefed an official from the Chinese Embassy on Labor green jobs initiatives.
ILAB	Briefing for Confederation of Venezuelan Labor Washington, DC liaison	Labor staff briefed the liaison on Labor green jobs efforts.
OSHA	Small Business Forum: "Green Jobs: Safety & Health Outlook for Workers and Small Employers"	A forum on OSHA's green jobs efforts and workplace hazards associated with green jobs.

Focus Area: Publicity and outreach		
Office	Name of effort	Reported description
OSHA	Presentation at the Center for Polyurethane Industry Annual Conference	Presentation – "What You Need to Know About the Safe Use of Spray Polyurethane Foam (SPF) Briefing on Spray Polyurethane Foam"
OSHA	Participate in NIOSH Going Green Workshop	OSHA Team attended as participating partner and Assistant Secretary spoke.
OSHA	US/EU Conference Brussels	OSHA co-chaired the topic, "OSH in Green Economy" for the conference on behalf of the United States. OSHA led the discussions and wrote the accompanying white paper.
OSHA	Green Jobs, Good Jobs National Conference	OSHA senior staff made presentations at conference on hazards of green jobs.
Focus Area: Publicity and outreach		
Office	Name of effort	Reported description
OSHA	Green Jobs, Good Jobs, Regional Conferences	OSHA personnel made presentations in Atlanta, GA; Los Angeles, CA; Philadelphia, PA; and Detroit, MI.
OSHA	Trilateral Roundtable: With Canada and European Commission	OSHA participated in: The Employment Dimension of the Transition to a Green Economy". The event brought together experts from government, trade unions, industry, and other stakeholders to exchange information, best practices, and ideas on preparing workers and employers to meet the increasingly complex skill demands of this transition. OSHA made a presentation on Green Jobs hazards.
OSHA	OSHA Alliance Program Construction Roundtable	Roundtable has received presentations from CPWR, NIOSH and Department of Commerce on green jobs within the construction industry.
OSHA	CleanMed Conference on Worker Safety, Patient Safety, Environmental Safety	First Annual Research Exchange on Advancing Patient, Worker and Environmental Safety and Sustainability in the Health Care Sector. OSHA presentation on focus on green jobs in relation to the healthcare industry. The audience was mainly healthcare workers, employers and researchers.
OSHA	OSHA – NIOSH Infosheet & Posters: Protecting Workers Who Use Cleaning Chemical	Provides information to employers on practices to help keep workers safe when working with cleaning chemicals, including green cleaning products. Theposters are available in English, Chinese, Tagalog and Spanish. The poster includes a section devoted to Green Cleaners.

(Continued)

Focus Area: Publicity and outreach		
Office	Name of effort	Reported description
OSHA	Presentation at American Industrial Hygiene Conference & Exposition	Topic: Making Green Jobs Good Jobs – We All Want To, So What is OSHA Doing to Make it Happen?
WB	Women and Green Jobs Roundtables	Discussions at over 30 U.S. locations involving business and community leaders regarding emerging employment opportunities in green job fields.
WB	Why Green is Your Color, outreach materials	Posters, mobile marketing displays, postcards, flash drives.
Focus Area: Technical assistance		
Office	Name of effort	Reported description
ETA	Green Jobs Community of Practice (CoP) on Workforce3One	ETA designed the Green jobs CoP to serve as a platform for workforce professionals and green job thought leaders to discuss and share promising practices, to create partnerships for green job workforce solutions, and to leverage Recovery Act investments. Specifically, the Green Jobs CoP was designed to provide an interactive platform for providing technical assistance through webinars, discussion boards, blogs and other online resources to workforce professionals, particularly those at the state and workforce investment board levels as well as green jobs grantees (including recipients of upcoming Solicitation for Grant Applications).
Focus Area: Training and supportive services		
Office	Name of effort	Reported description
ETA	YouthBuild	The YouthBuild program targets out-of-school youth ages 16 to 24 and provides them with an alternative education pathway to a high school diploma or GED. Most YouthBuild programs have incorporated green building into theirconstruction training. As part of this training, participants learn about environmental issues that affect their communities and how they can provide leadership in this area.

Focus Area: Training and supportive services		
Office	Name of effort	Reported description
OSHA	Blue-Green Alliance Conference	Attended the Blue-Green Alliance conference
Focus Area: Training and supportive services		
Office	Name of effort	Reported description
VETS	Homeless Veterans' Reintegration Program	The purpose of this program is to expedite the reintegration of homeless veterans into the labor force. These grants are intended to address two objectives: to provide services to assist in reintegrating homeless veterans intomeaningful employment within the labor force, and to stimulate the development of effective service delivery systems that will address the complex problems facing homeless veterans. The programs' technical assistance guide refers to collecting data on green jobs participants.
WB	Why Green is Your Color, Web-based training	Web-based training to help women find and succeed in green jobs.
WB	Green Jobs for Women Pilot Training Projects	Pilot training projects designed to prepare women to enter high-growth, high-demand green jobs.

Source: Responses provided by Labor.

End Notes

[1] See 29 U.S.C. § 2916(e)(1)(B) for a list of these industries.

[2] The Green Jobs Act of 2007, which amended the Workforce Investment Act, was enacted as part of the Energy Independence and Security Act of 2007 (EISA). Pub. L. No. 110-140, 121 Stat. 1492, 1748. EISA concerned, among other things, energy initiatives such as increasing fuel economy standards, increasing the production of biofuels, and improving appliance and lighting standards.

[3] Pub. L. No. 111-5, 123 Stat. 115, 173.

[4] Pub. L. No. 111-5, § 901, 123 Stat. 115, 191. For a list of our Recovery Act-related products, see http://www.gao.gov/recovery.

[5] For the purposes of this report we refer to organizational entities within Labor as offices. These entities could include agencies, administrations, bureaus, centers, and divisions. Specifically, the 14 offices we identified are: Occupational Safety and Health Administration (OSHA), Mine Safety and Health Administration (MSHA), Women's Bureau (WB), Employment and Training Administration (ETA), Veterans' Employment and Training Services (VETS), Office of the Assistant Secretary for Policy (OASP), Bureau of International Labor Affairs (ILAB), Bureau of Labor Statistics (BLS), Center for Faith-Based and Neighborhood Partnerships (CFBNP), Office of Federal Contract Compliance

Programs (OFCCP), Wage and Hour Division (WHD), Office of Workers' Compensation Programs (OWCP), Office of Disability Employment Policy (ODEP), and the Office of Public Affairs (OPA).

[6] For the 14 directly-funded green efforts in our scope, we received 13 completed questionnaires and one partially completed questionnaire. We also identified 3 additional directly-funded efforts, for a total of 17 efforts. For the second set of green jobs efforts where funds were not specifically appropriated or allocated for green jobs activities, but where the office sought to incorporate green elements into either an existing program or ongoing activity, we received completed questionnaires for 46 efforts in our scope and we did not survey Labor officials for 2 of the 48 efforts later identified. See appendix I for more information.

[7] We visited grantees in California, Illinois, Minnesota, and Pennsylvania, and interviewed grantees in Connecticut and Louisiana by phone.

[8] Pub. L. No. 110-140, Title X, 121 Stat. 1492, 1748.

[9] U.S. Department of Labor, Office of Inspector General, *Recovery Act: Employment and Training Administration Grant Issuance and Monitoring Policies and Procedures for Discretionary Grants Including Green Jobs Are Comprehensive but Funding Challenges Threaten the Quality of Future Monitoring Activities*, 18-10-013-03-390 (Sept. 30, 2010); *Recovery Act: Slow Pace Placing Workers Into Jobs Jeopardizes Employment Goals of the Green Jobs Program*, 18-11-004-03-390 (Sept. 30, 2011); and *Recovery Act: Green Jobs Program Reports Limited Success in Meeting Employment and Retention Goals as of June 30, 2012*, 18-13-001-03-390 (Oct. 25, 2012).

[10] These seven industries are (1) energy efficient building, construction, and retrofits; (2) renewable electric power; (3) energy efficient and advanced drive train vehicles; (4) biofuels; (5) deconstruction and materials use industries; (6) energy efficiency assessment serving the residential, commercial, or industrial sectors; and (7) manufacturers that produce sustainable products using environmentally sustainable processes and materials.

[11] The Occupational Information Network (O*NET) is overseen by ETA and is the nation's primary source of occupational information. Central to the project is the O*NET database, which contains information on hundreds of standardized and occupation-specific descriptors. O*NET is being developed under the sponsorship of the U.S. Department of Labor's Employment and Training Administration through a grant to the North Carolina Department of Commerce. For more information on O*NET's green economy activities, see http://www.onetcenter.org/green.html.

[12] According to BLS's definition, green jobs are either (1) jobs in businesses that produce goods or provide services that benefit the environment or conserve natural resources, or (2) jobs in which workers' duties involve making their establishment's production processes more environmentally friendly or use fewer natural resources.

[13] Focus areas generally fell into 1 of 7 categories provided by GAO: (1) information materials; (2) publicity and outreach; (3) technical assistance; (4) partnerships; (5) training and supportive services; (6) data collection and reporting; and (7) capacity building.

[14] We requested information from 14 different offices across Labor. These 14 offices were selected based on the likelihood of their administering a green jobs effort or program.

[15] Labor provided extensions to the original period of performance for selected grantees in each of the five Recovery Act green jobs grant programs (see fig. 3).

[16] According to Labor officials, the BLS Green Jobs initiative was not connected to the Recovery Act in any way.

[17] This grant is included in the $89 million identified by Labor as allocated or appropriated for green jobs activities in addition to the Recovery Act-funded programs (see fig. 2).

[18] ETA refers to this time as the period of performance.

[19] In total, ETA awarded 189 green jobs grants with Recovery Act funding, of which 97 were primarily focused on training. In addition, ETA awarded 6 GJIF training grants for green jobs training. The analysis in this section pertains to the 103 training specific green jobs grants awarded through the 3 Recovery Act-funded programs primarily focused on training (SESP, POP, and ETP) and the GJIF program, which was funded as a pilot and demonstration project under WIA.

[20] Over half of grantees provided training through a network of local affiliates or sub-affiliates, sometimes within a single state or located in multiple states. Labor officials could not provide the exact number or location of such sub-grantees for the three Recovery Act-funded green jobs training programs, so this analysis includes only the location of primary grant recipients.

[21] As defined by the U.S. Census Bureau.

[22] Labor provides oversight of sub-grantees through its monitoring of primary grantees, which includes ensuring that primary grantees monitor their sub-grantees and document monitoring results.

[23] Ensuring that green jobs would be accessible to women and minorities was identified as a significant challenge in the Vice President's 2009 Middle Class Task Force report, *Green Jobs: a Pathway to a Strong Middle Class.*

[24] While there are 48 efforts in total, we did not survey Labor officials for 2 of the 48 efforts listed in appendix III.

[25] The number of grantees interviewed and training programs examined totaled 11. Therefore, for the remainder of this section, only the relevant count of grantees or programs will be presented within parentheses and it can be assumed that the count is out of 11. It can also be assumed that the qualifier "all" equates to 11.

[26] BLS counts a job as being green if it is held by a worker who spends over half their time involved in green technologies and practices.

[27] Labor has asserted the importance of the involvement of local employers in the alignment of employment training programs with jobs. For example, WIA requires that a majority of local workforce investment boards be representatives of local businesses. GAO has also reported on the importance of linking the workforce investment system with employers to meet local labor market needs. See GAO, *Workforce Investment Act: Innovative Collaborations between Workforce Boards and Employers Helped Meet Local Needs*, GAO-12-97 (Washington, D.C. Jan. 19, 2012).

[28] Labor officials told us that these time frames are based on grantee performance reporting requirements.

[29] Collectively for these grantees, the number of program participants enrolled was 104 percent of the grantees' aggregate target level, and the number of program participants completing training was 106 percent of the grantees' aggregate target level.

[30] This analysis of reported outcome data only includes grantees in two of the three Recovery-Act funded green jobs training programs because final outcome data were not available for any grantees of the third program. Specifically, all grantees that had reported final outcome data as of December 31, 2012, were either Pathways Out of Poverty grantees or Energy Training Partnership grantees. Almost all grantees for which final outcome data had not yet been reported were grant recipients of the State Energy Sector and Partnership and Training (SESP) grants. Employment-placement outcomes and earnings might vary systematically across grant programs due to differences in the populations targeted by the grants. For example, Pathways Out of Poverty grantees served unemployed individuals, high school

dropouts, and other disadvantaged individuals within areas of high poverty. In contrast, under SESP grants, various target populations were given priority to receive training, such as workers impacted by national energy and environmental policy, veterans, and unemployed individuals, among other groups. Other individuals, such as entry-level and incumbent workers were also eligible to be served through the SESP program.

[31] For their data reliability assessment, Labor's OIG performed a file review of outcome data for eight randomly selected grantees of the Recovery Act programs. They found that 24 percent of entered employment outcomes were not supported by adequate documentation, and that 33 percent of training-related entered employment outcomes were not supported by adequate documentation. Most of the documentation problems were found with three of the eight sampled grantees.

[32] For a complete description of our data reliability assessment process, see appendix I.

[33] GAO previously reported that UI wage records provide a fairly consistent national view of the performance of WIA. The report noted that most of the outcome data reported in a given program year actually reflect participants who left the program during the prior year, which limits their usefulness for gauging current program performance. GAO has not recently assessed how Labor calculates entered employment rates, job retention rates, or average earnings using these data. See: GAO, *Workforce Investment Act: States and Local Areas Have Developed Strategies to Assess Performance, but Labor Could Do More to Help,* GAO-04-657, (Washington, D.C.: June 1, 2004).

[34] According to Labor, their most recent analyses of Recovery Act green jobs training program outcomes using UI wage data are based on data as of December 31, 2012. The data contain employment information for program participants who had exited training during the one-year period ending March 31, 2012. The data contain job retention and earnings information for program participants who had exited training during the one-year period ending September 30, 2011.

[35] According to Labor, their most recent analyses of the GJIF green jobs training program outcomes using UI wage data are based on data as of December 31, 2012. The data contain employment information for program participants who had exited training during the one-year period ending March 31, 2012. Also according to Labor, relevant UI wage data to generate job retention rates or average earnings for participants of the GJIF program are not yet available.

[36] All 11 grantees we interviewed were of the opinion there is a national movement toward energy efficiency due to economic changes, whereby industries or singular businesses are attempting to be globally competitive partly by reducing energy costs. Ten of the grantees were also of the opinion that such a national movement is also being driven by policy changes, such as state energy goals, and by cultural changes, such as public concern about global warming or public desire for U.S. energy independence. One grantee was not of the opinion that such a national movement was being driven by policy or cultural changes.

[37] The three surveys launched by BLS are (1) the 2010 and 2011 Green Goods and Services (GGS) surveys; (2) the Green Technologies and Practices (GTP) survey; and (3) the Occupational Employment and Wages in Green Goods and Services (GGS-OCC) program.

[38] U.S. Department of Labor, *Green Jobs and Health Care Implementation Study Final Report* (Sept. 20, 2012), noted that using pre-existing partner relationships contributed to the efficiency of program implementation. Further, the report suggested that requiring future grant administrators to identify partners with whom they have had positive preexisting relationships would likely limit the amount of up-front administrative time spent working with partners—especially for grants with shorter periods of performance.

[39] The ETP, POP, and SESP grants. The evaluation also examined the implementation of the Health Care Sector and and other High Growth and Emerging Industries grant, which did not specifically target green jobs.

[40] Ted Shen, Eileen Poe-Yamagata, Alan Dodkowitz, et. al., *Green Jobs and Healthcare Implementation Study: Interim Report,* ETAOP 2012-07 (IMPAQ International: May 7, 2012). Please also see: Michael P. Kirsch, Carolyn Corea, Ted Shen, et. al., *Green Jobs and Healthcare Implementation Study Final Report*, ETAOP 2013-16 (IMPAQ International: Mar. 27, 2013).

[41] Diane Herz, Elizabeth Laird, Angus Hague, et. al., *Investing in Labor–Market Information (LMI): Insights from the Recovery Act LMI Grants Final Report,* ETAOP 2013- 05 (Mathematica Policy Research, Inc.: Jan. 15, 2013). See also: Jillian Berk, Elizabeth Laird, and Brittany English, *Investing in Labor–Market Information (LMI): A Summary of the State LMI Improvement Grants Final Report,* ETAOP 2013-06. (Mathematica Policy Research, Inc. Jan. 15, 2013). Finally, see: Elizabeth Laird, Brittany English, Jillian Berk, et.al., *Practitioner Briefs for the Evaluation of the Labor Market Information (LMI) Improvement Grants,* ETAOP 2013-17 (Mathematica Policy Research, Inc. Jan. 15, 2013).

[42] The evaluation also assesses impacts of selected grantees of another Recovery Act-funded training program, the Health Care Sector and Other High Growth and Emerging Industries grant.

[43] Specifically, the National Governors Association, Corporation for a Skilled Workforce, and Collaborative Economics form the TA Partnership.

[44] GAO, *Internal Control: Standards for Internal Control in the Federal Government,* GAO/AIMD-00-21.3.1 (Washington, D.C.: November 1999).

[45] U.S. Department of Labor, Employment and Training Administration, *Core Monitoring Guide* (Washington D.C.: April 2005).

[46] Federal project officers have overall responsibility for monitoring the conduct and progress of grantees, including conducting on-site visits. Specifically, they are responsible for collaborating with the grantees—both in the planning and implementation of the program and in the evaluation of activities.

[47] Green jobs grantees were required to submit two kinds of quarterly performance reports, one that aggregated performance data and the other that described the status of program implementation in narrative form.

[48] While the risk level is primarily driven by the grantee's overall performance and financial performance, other factors also affect risk, such as the level of staff turnover at the grantee's organization, loss of key grantee personnel, or other staff issues, according to Labor officials.

[49] ETA officials noted that the green jobs training programs were funded through grants and not through contracts and that this is statement is true across programs of federal assistance.

[50] See GAO, *Grants to State and Local Governments: An Overview of Funding Levels and Selected Challenges* GAO-12-1016 (Washington, D.C.: September 25, 2012),.24 and U.S. Department of Justice Office of the Inspector General, *Improving the Grant Management Process* (February 2009).

End Notes for Appendix I

[1] For the purposes of this report we refer to organizational entities within Labor as offices. These entities could include agencies, administrations, bureaus, centers, and divisions.

[2] We did not receive a completed survey for one of the efforts, but we did receive funding information for the effort that was requested in the survey in an e-mail from Labor on April 3, 2013.

INDEX

A

B

C

D

E

F

G

H

I

J

L

M

N

O

P

R